Cambridge Elements ≡

Elements in the Philosophy of Biology
edited by
Grant Ramsey
KU Leuven
Michael Ruse
Florida State University

THE BIOLOGY OF ART

Richard A. Richards
University of Alabama

CAMBRIDGE
UNIVERSITY PRESS

University Printing House, Cambridge CB2 8BS, United Kingdom

One Liberty Plaza, 20th Floor, New York, NY 10006, USA

477 Williamstown Road, Port Melbourne, VIC 3207, Australia

314–321, 3rd Floor, Plot 3, Splendor Forum, Jasola District Centre,
New Delhi – 110025, India

79 Anson Road, #06–04/06, Singapore 079906

Cambridge University Press is part of the University of Cambridge.

It furthers the University's mission by disseminating knowledge in the pursuit of
education, learning, and research at the highest international levels of excellence.

www.cambridge.org
Information on this title: www.cambridge.org/9781108727846
DOI: 10.1017/9781108672078

First published 2019

A catalogue record for this publication is available from the British Library.

ISBN 978-1-108-72784-6 Paperback
ISSN 2515-1126 (online)
ISSN 2515-1118 (print)

Cambridge University Press has no responsibility for the persistence or accuracy of
URLs for external or third-party internet websites referred to in this publication
and does not guarantee that any content on such websites is, or will remain,
accurate or appropriate.

The Biology of Art

Elements in the Philosophy of Biology

DOI: 10.1017/9781108672078
First published online: April 2019

Richard A. Richards
University of Alabama

Abstract: Biological accounts of art typically start with evolutionary, psychological or neurobiological theories. These approaches might be able to explain many of the similarities we see in art behaviors within and across human populations, but they don't obviously explain the differences we also see. Nor do they give us guidance on how we should engage with art, or on the conceptual basis for art. A more comprehensive framework, based also on the ecology of art and how art behaviors are expressed in engineered niches, can help us better understand the full range of art behaviors, their normativity and their conceptual basis.

Keywords: philosophy, biology, art

ISBNs: 9781108727846 (PB), 9781108672078 (OC)
ISSNs: 2515-1126 (online), 2515-1118 (print)

Contents

1 What Is Art?

1.1 Introduction

The title of this Element, *The Biology of Art*, raises two questions: *What* is the biology of art? *Why* a biology of art? The next four sections are largely an answer to the first question. The answer to the second question, asking for justification of a biology of art, will obviously depend partly on the answer to the first. We might begin to answer, though, by noting the spectacular success of science in general, and of biology in particular. If through biological investigation, for instance, we can understand and cure diseases, why not use similar methods to understand art? Skeptics might answer, however, that art consists of cultural practices and traditions grounded on a conceptual background and can therefore only be explained using the resources of philosophy, art history, cultural anthropology and related disciplines. To see if this sort of skepticism is justified, we need to look at a full development of the biological approach. Only then will we see the range and depth of its explanatory power.

1.2 Defining *Art*

But first, what is art? If we are going to give a biological account of *art*, we need to have some idea of what it is that we are explaining. We might take this question to be asking only for a definition of the term "art" when we say that some object or activity is a work of art. Such a definition would then just be telling us how that term is used. But if that is all we are after, a good dictionary will answer our question and this inquiry will be done. Instead, we might be asking about the idea or concept of *art* that lies behind the use of the term "art," and that *best* reflects, clarifies and organizes our thinking and practices. If this is our project, to answer the question "what is *art*?" we could adopt a method with a long history, Plato's "elenchus," which begins an inquiry with a definition of an idea or a concept, and then tests that definition by its application to the world. In his *Republic*, for instance, Plato asks the question "what is justice?" and then addresses a variety of answers. Each answer – some definition of justice – is tested by its application to the world. If a definition implies incorrectly that some truly unjust actions are just, or that some truly just actions are unjust, then that definition is modified to accommodate these counterexamples. The new definition is then tested to find counterexamples. This process is repeated until the true, clarified definition is found.

Using this method to arrive at a satisfactory definition of *art* (as idea or concept), we start with some definition, then apply that definition to see if it correctly identifies all and only those things that are truly works of art. If it doesn't, we revise the definition to better identify works of art. The history of philosophical thinking about art is filled with what could be regarded as definitions in this sense. We can begin with the idea of *mimesis*, usually understood as imitation or representation, developed by Plato in books 2, 3 and 10 of his *Republic* (Plato 1997) (Janaway 2002), and by Aristotle in his *Poetics* (Aristotle 1987, 1448b6). In drama, for instance, an actor imitates the actions or emotions of another person, and the plot of the play represents events. But as a definition of art in the modern sense, this will not do. Not all things that seem to be art engage in this kind of imitation or representation. Abstract paintings and instrumental music are just two obvious counterexamples.

Perhaps the distinctive functioning of art is not to imitate or to represent, but to produce a certain kind of pleasure. After all, we enjoy listening to music, looking at paintings, reading poetry and watching drama. This view, typically attributed to David Hume (Goldman 2012), is not all wrong, but it is easy to find counterexamples. Some plays and films are disturbing, and instead generate anxiety, angst or even disgust. Nonetheless, they still seem to be legitimate instances of art.

We might instead associate art with imagination. Immanuel Kant identified the experience of art with the stimulation of the disinterested, harmonious free play of the imagination and understanding (Kant 1978; Schellekens 2012). Kendall Walton (1990) has more recently identified a distinctive role for the imagination in art, as the foundation for make-believe. But as in the case of mimesis and pleasure, some commonly accepted instances of art seem to be counterexamples. It isn't clear that Bach's concertos stimulate the imagination, at least in the way a landscape painting might.

Perhaps instead we should conceive of *art* as expression. According to Leo Tolstoy (1899), true art expresses the artist's emotion, and then stimulates that same emotion in the audience. Benedetto Croce (1938) and R. G. Collingwood (1945) have argued for similar, but less personal accounts of expression, in that the emotion expressed need not be an emotion actually experienced by the artist (Kemp 2012). While some art (music in particular) is clearly expressive in some fashion or other, it isn't clear that *all* art is expressive in these ways. Piet Mondrian's geometric paintings, for instance, don't seem expressive in the sense advocated by Tolstoy, Croce and Collingwood.

Perhaps art should instead be understood as the production of a distinctly aesthetic experience. The function of a painting might be to give an aesthetic visual experience, and the function of music might be to give an aesthetic aural experience. Clive Bell (1914) postulated a distinctive aesthetic emotion, stimulated by "significant form," based on the formal qualities of an artwork such as the color, line and composition of a painting. Monroe Beardsley (1982), like Bell, started with the idea that art seems to stimulate the senses in distinctive ways. We often have distinctive perceptual experiences when we experience art that we don't seem to have otherwise. But here, as with other theories of art, there are apparent counterexamples. Duchamp's "readymades" – his snow shovel, bottle rack and urinal – don't obviously stimulate these aesthetic experiences in the ways suggested, yet they are now typically regarded as art.

1.3 Procedural Criteria

The distinctive features of art identified in the theories just surveyed can broadly be described as *functional*, in the sense that art is identified with the particular (and valuable) things it can do. But we might instead identify art by how it is produced. Stephen Davies calls these "procedural" definitions of art, and distinguishes them from the functional (S. Davies 1991, 2002). On this approach an object can be a work of art by virtue of being created within a particular art tradition or institution. A new ballet produced by the Bolshoi ballet company, for instance, will be a ballet because it is produced by a *ballet* institution, and within the *ballet* tradition. More generally, it will be *art* on the same grounds – the history of the Bolshoi ballet institution in producing ballet instances of art. Arthur Danto and George Dickie advocate these sorts of procedural criteria for identifying art. In their terms, for something to count as *art*, it must be a product of the "artworld." For Danto (1964), the emphasis is on art theory and the historical tradition. (See also Levinson 1979, 2002, and N. Carroll 1988, 1990.) For Dickie (1974), the emphasis is on the role of institutions and established practices in doing things.

As with the other criteria, this approach captures something important about what counts as *art* – the social and historical context in which it is produced. But also, as with the other criteria, there seem to be counterexamples. Prehistoric cave paintings may look like art and function like art in producing pleasure or an aesthetic experience, but the first cave paintings could not possibly have been produced as part of an art tradition or through an art institution. There were none. Therefore they cannot be art in a straightforward procedural way. But even if the traditions and institutions are important to the modern practice of art, do we really want to say that *in principle* no one can ever have produced art except within an established art tradition or institutional context? We might agree here with Stephen Davies (2015), who thinks an adequate definition of art must account for the cases of things that look like art – cave paintings, for instance – but that weren't produced within an art tradition or institution. Otherwise, if we don't know about the relevant art tradition or institutional context, as in the case of much prehistoric cave art, we simply cannot know whether something is a work of art or not. This is surely problematic.

These procedural approaches must also confront two related problems. First, they push questions about the definition of art back to the questions "what makes a tradition an *art* tradition?" and "what makes an institution an *art* institution?" If we need to know whether something is a product of an *art* tradition or institution to identify it as art, we need to know what makes something that kind of tradition or institution, and not some other kind of tradition or institution. How, for instance, do we distinguish *art* traditions and institutions from *craft* traditions or institutions? Perhaps we can identify something as an *art* tradition or institution *only* because we have some prior way of identifying something as a work of art (Stecker 1996)!

Second is the *artworld relativity* problem. If there are multiple artworlds – multiple art traditions, institutions and practices – then one artworld might regard something as a work of art, while another might not (S. Davies 2002, 174–175). This might be the case with some of the more controversial instances of recent art, Duchamp's readymade "Fountain" urinal, for instance. Elite traditions, institutions and practices might treat it as art, whereas folk traditions, institutions and practices might not. And contemporary artworlds might regard it as art, whereas 18th-century artworlds may not. We might be tempted to understand our own preferred artworld as authoritative, but it isn't obvious why we should. While these problems may not necessarily refute procedural approaches, they do present serious worries.

Perhaps a hybrid approach that combines functional and procedural criteria can avoid some of these problems with the simple procedural and functional criteria. Stephen Davies argues that there is more than one way something can qualify as art: a) if it exhibits skill in serving some aesthetic goal; b) if it falls under an art genre or form within a recognized art tradition; or c) if it is intended by its maker to be art, who does what is appropriate and necessary within the historical context to realize that intention (S. Davies 2015, 377–378; see also Stecker 1997). This solves one problem in that it may better reflect our intuitions that some works produced outside of art traditions and institutions, such as the first cave paintings, really can be art, but that art traditions and institutions also sometimes play a role in establishing what counts as art. But since none of these conditions is necessary for something to count as art, we might conclude that there are then three kinds of art, each satisfying one of the criteria – art as aesthetic object, art as tradition and art as intentional action. If so, there is no single thing that is *art*.

1.4 Classical versus Cluster Definitions

A careful evaluation of these claims about the defining features of art is beyond the scope of this Element. It may be that some of the counterexamples can be explained away, or that some definitions can be refined to avoid some of the problems. But what this brief survey seems to reveal is that while there are functional features that are distinctive of some art, those features are also lacking in other recognized instances of art. Therefore these features cannot be *necessary* for something to be art. It cannot be that art *must* represent something, express an emotion, produce pleasure, stimulate the imagination or produce an aesthetic experience. Moreover, these features don't seem *sufficient* to make something art. Mere representation doesn't make something art, as some road signs make clear. Nor does mere expression make something art, as crying, laughing and cursing are all expressive. And production of an aesthetic experience is also insufficient in that an aesthetic experience might be produced by a beautiful person, a colorful flower or bird, a landscape, an automobile or a piece of driftwood.

The procedural criteria faces similar problems. If it is at least plausible to argue that the first cave art is still art, even though there were no art traditions or art institutions at the time, then the procedures based on tradition and institution cannot be *necessary*. And if it is at least plausible to argue that some action or object produced within an art tradition, or with the approval of an art institution, is not actually art, then the procedures don't seem sufficient. We might doubt that Duchamp's readymades really are art, even though they are institutionally regarded as works of art. Do we really want to say that anything is art *solely* because it is called "art" by some art institution or authority? As important as the traditions and institutions may be, there are legitimate worries that they can provide necessary and sufficient conditions for being a work of art.

But perhaps we should think about defining art not in terms of necessary and sufficient conditions but as a cluster of conditions. Dennis Dutton (2010) has argued that there is no set of necessary and sufficient conditions. Instead there are 12 *relevant* criteria associated with art and the production and experience of art: 1) art produces pleasure; 2) art exhibits skill and virtuosity; 3) art exhibits style; 4) art exhibits novelty and creativity; 5) art has a tradition of criticism; 6) art involves representation; 7) art has special focus; 8) art is expressive individuality; 9) art has emotional saturation; 10) art presents intellectual challenge; 11) art has traditions and institutions; 12) art involves imaginative experience (Dutton 2010, 59). On this approach, something is art if it has a sufficient subset of these features. And different subsets may do the job. Something might only have criteria 2, 3, 4, 8 and 11, for instance, and still be a work of art. Or it might have criteria 1, 5, 7 and 9. This cluster approach seems to avoid the obvious problem that no set of defining conditions seems really necessary or sufficient. Perhaps *art* is just one of those fuzzy categories that cannot be neatly defined in this way.

While there is something attractive about this cluster approach, it cannot obviously answer some questions. First, how and why should we come up with precisely that list? Each art object has other properties or features that might be relevant. For instance, paintings are physical objects. Why not include "art is a physical object" on the list? Dutton doesn't say how this list is generated, other than it is "what we already know about the arts." Second, how many of the 12 features must be present for something to be a work of art? Five? Seven? Nine? And on what grounds do we say that a particular subset of these criteria is sufficient? Dutton admits there is no formula for deciding, but seems to think that this is nonetheless a useful guide for identifying central cases and assessing marginal cases (Dutton 2010, 60–62). Third, are some of these features more important than others? Are novelty and creativity more important than skill? The cluster approach may reflect our intuitions, but it cannot be the final word. It leaves too many questions unanswered.

1.5 Theoretical Problems

The failure so far of the functional and procedural art criteria is practical, in that they cannot be applied to pick out all and only true works of art. But there are also theoretical reasons why these art criteria might fail to give us an adequate definition. First, as Paul Kristeller noted in a widely cited article from 1951, the modern way of classifying things as *art* seems to have its origin in 18th-century European thinking. According to Kristeller, the "modern system of the arts" has at its nucleus the five major arts – painting, sculpture, architecture, music and poetry – and sometimes includes gardening, engraving, the decorative arts, dance theatre and opera. But before the 18th century, there was no tradition of grouping these arts together into a single kind of unified activity, and comparing them on the basis of common principles – treating them all as instances of *art*. Rather each of the arts was conceived independently with most of the discussion about technical principles in practicing that art. We see this clearly in Aristotle's *Poetics*, which at its core seems to be a discussion of the technical principles governing poetry. Moreover, the term "art" itself may be traced back to the Greek term "techne" and its Latin equivalent, "ars," both applied to all kinds of activities, including those we would now think of as crafts and sciences, and referring to the techniques or methods of doing these things (Kristeller 1951, 479–498).

The grouping of these activities that later came to be seen as art in the modern sense varied over time. At one time, according to Kristeller, music was a category that included poetry, with dance just as an element of poetry. Sometimes music was combined with math, geometry or astronomy. The visual arts only came to be included with the other arts (in the modern sense) at the end of the 14th century, while the sciences were only becoming divorced from the arts in the 17th century (Kristeller 1951, 501–525).

The point here is that our thinking about what we now call "art" is a relatively new way of thinking. Perhaps we shouldn't assume that there is some static, unchanging concept *art*, and some static, unchanging set of things that we can identify as works of *art*, and against which we can test our definitions. What counts as *art* at one time may not count as so at other times. If so, at what time should we take the *art* classification to be authoritative? Should we adopt the 1951 classification in Kristeller's analysis? Or should we adopt a 21st-century classification? At minimum, this suggests that when we ask what art is, we need to index it to a particular time. And any definition of the term or concept will *at best* be limited to a particular time.

Moreover, if we recognize that art practices are embedded in culture, then it is not clear there is a single cross-cultural thing *art*. Cultural anthropologist Clifford Geertz argues that cultural practices are based on cultural meanings, and that because each culture interprets the world differently, and understands its practices differently, the so-called arts of various cultures are not the same kind of thing (Geertz 1976, 1475–1476). The colorful and highly decorated Abelam yam masks, for instance, function in competitive exchange, harvest ceremonies and festivals, and are imbued with spiritual meanings (Scaglion 1993). The colorful Impressionist paintings in Europe and the United States, on the other hand, function within a system of art museums and commercially produced posters. This suggests that on cultural grounds there can be no single, universal thing *art*. If so, the definition project seems to fail.

1.6 Theory and Art

So far, the functional and procedural approaches have been superficial, in that they have focused on the obvious and manifest ways that artworks do things and function, and on the ways that they become artworks. It is obvious that drama often represents. We can just *see* the representation in plays. It is obvious that music often expresses. We can just *hear* the expression in music. And it is obvious that Impressionist paintings produce aesthetic experiences. We can just *see* the Impressionistic visual effects. It is also obvious that certain institutions call some things art, and not others, and have done so within art traditions.

The approaches outlined here are also naturalistic and empirical in the broad sense that they generally do not appeal to supernatural entities or processes, and are based on observation (looking at plays and paintings and listening to music). This is an approach that Francis Bacon might have advocated. He is famous (or infamous) for arguing that the correct approach to the inquiry into nature starts with observation unsullied by theory. We can go wrong in our inquiries, according to Bacon, by relying on the theories and dogmas of the past – what he called the "Idols of the Theatre" (Bacon 1960, Bk. I, XLIV). To avoid this bias, we should just look at the world, without assumption of any theory at all, classify what we observe, and then generalize. An investigation of the nature of heat, to take one of Bacon's examples, should begin by observing all things that have heat, and contrasting them with those things that lack heat. Bacon illustrated this approach by identifying 28 instances of heat, including the sun's rays, reflected sun's rays, flaming meteors, lightning, any flame and heated or boiling things (Bacon 1960, Bk. II).

Anyone who knows modern scientific thinking about heat will also know that Bacon's atheoretical approach did not win the day. Heat, as it is now conceived, is in terms of the average kinetic energy of particles, and cannot just be observed in the world. We could look at a lot of hot things and never generalize to some law about average kinetic energy, because we cannot simply and directly observe either the particles in motion or the average kinetic energy of these particles in the way we can observe temperature by looking at a thermometer. To understand heat phenomena, it is necessary to start with a theory – in this case the kinetic energy theory. Perhaps to understand art we need to do more than just observe artworks and activities, and then classify and generalize in the Baconian mode. Perhaps we need to start with some theory.

We can understand this theoretical approach by analogy with modern chemistry. We could begin to define *water*, for instance, by its manifest, exhibited properties. It is clear and transparent. It is wet. It satisfies thirst. It cleans things. It falls from the sky. It bubbles up from the ground. It is a solid when cold, and a gas when heated. But modern chemistry tells us that these exhibited traits are not what is crucial. What is crucial is molecular composition: Water is H_2O – one oxygen atom bound to two hydrogen atoms. Can *art* be given a similar theoretical definition? To answer this, we simply need to try. And we begin with a fully "naturalistic" approach.

1.7 Naturalism

Recently there has been an extensive debate about the nature and justification of naturalism in philosophy. While there are many versions of naturalism, often it is divided into two sets of commitments: metaphysical or ontological – about what exists – and methodological – about the best method for explaining phenomena. On the first, ontological commitment, naturalists, at a minimum, deny the existence of supernatural entities and processes. On the second, methodological commitment, naturalists tend to give special authority to scientific methods (Papineau 2015; Clark 2016). On one naturalistic approach, what we might call "scientific naturalism," philosophy should begin with what the best scientific theories tell us about the world (Clark 2016, 3–5). If so, we will then be relying on the substance of the science – what entities and processes it uses to explain phenomena – and scientific methods – how it comes to accept these entities and processes. The advantage of this approach is obvious. Just as science has given us a uniquely successful way of understanding the world, it may give us a successful way of understanding art.

Suppose we start with a scientific and theoretical approach. With which scientific theories should we begin? A pragmatic stance suggests that the answer to this question depends on what we are investigating and what we want to know. If we are interested in the visual perception of color in Impressionist paintings, for instance, we might begin with theories in visual psychophysics. If we are interested in the heredity and development of musical abilities, we might begin with theories related to genetics and epigenetics. But this approach is also objective in that the theories we use can be assessed for how well they are supported by evidence, how coherent they are and whether they are consistent with other good theories. If so, we have objective grounds for evaluating our starting point. The remainder of this Element is naturalistic in this sense, in that it begins with our best scientific theories as they inform us about the nature of art.

1.8 Conclusion

The failure to arrive at a good, clear account of the nature of art leaves us without a good, clear account of what this Element is about. We don't yet know what a scientific naturalistic account of art will look like because we don't yet know what counts as *art*. My proposal is that we start with a vague and tentative understanding of art based on what seems to be central and unproblematic cases of music, dance, storytelling and the visual arts. Perhaps after we have a more complete biology of art, we can say something more precise. It may be that ultimately we have to revise our thinking about the category of *art*. I address that possibility in the final section. This initial failure to give a single, unified, satisfactory account of art also suggests that it may be misleading to speak and write about "art" rather than "the arts." Whether or not that is correct,

I continue to write of art singularly as "art," but sometimes pluralistically in terms of "the arts." I do not intend either locution to imply anything about the ultimate nature of art.

At the beginning of this section, I suggested the title of this Element, *The Biology of Art*, raises two questions: "What is art?" and "Why a biology of art?" There is also a third question: What is biology? This last question is more fully answered in Sections 3, 4 and 5, but some preliminary comments are in order. We should obviously include the core biological commitments of evolutionary theory, genetics and epigenetics in our biology. The account here also includes psychology and neurobiology. These may not be thought to be a proper part of biology as it is usually organized in our universities. But if we begin with evolution, and recognize the continuity between humans and other organisms, then shouldn't human psychology be as much a part of biology as is animal psychology? Finally, as we see in what follows, ecology will also be part of the full framework. That is an addition that is long overdue, and it adds great explanatory power to the biological framework.

Before we look at the full biological framework, though, in the next section, we take a look at some recent efforts to naturalize our thinking about art, based on evolutionary theory, psychology and neurological functioning, and some philosophical objections to these efforts. In Section 3, we begin to understand the full biological framework by looking more closely at evolutionary theories of art. Section 4 returns to the psychology and neurobiology of art. Section 5 introduces the ecology of art, and niches in engineering. Section 6 starts with a general theory of the value of art and returns to the topic of this section – the nature of art – but with the benefit of the full biological framework.

2 Naturalism and Its Discontents

2.1 Introduction

The basic approach advocated at the end of Section 1 is a scientific naturalism that begins with our best scientific theories, and is based on a biological framework. One premise of this Element is that the biological framework has not yet been fully explored or developed. But some parts have been explored, in particular the evolution, psychology and neurobiology of art. As we see in what follows, these efforts have also prompted philosophical criticisms. First, while scientific approaches can tell us how we actually engage with art, they cannot answer the important philosophical questions about what constitutes the *proper* conception, experience and value of art. Second, scientific approaches do not and cannot study the essential, conceptual basis of art. These criticisms suggest that a scientific approach to art has very limited philosophical value.

In this section, we first look at some of these recent scientific approaches to art, and then at the challenges posed by these criticisms. In later sections, we see how a more complete scientific naturalism can respond.

2.2 Evolutionary Theories

In his *Origin of Species*, first published in 1859, Charles Darwin argued that all species evolved from common ancestry, with change occurring mostly, but not exclusively, on the basis of natural selection. While he hardly mentioned human evolution in his *Origin*, he discussed it extensively in the two volumes of his 1871 *On the Descent of Man and Selection in Relation to Sex*. In this later work and elsewhere, Darwin used two main explanatory strategies for understanding humans and their behaviors, based on natural *survival* selection, or sexual selection. Recent evolutionary accounts of art now generally start with one of these two selection-based explanatory strategies. First are the theories that explain art behaviors through survival selection. The idea is that the making and experiencing of art conferred a survival advantage to our ancestors, so that those individuals who engaged in art behaviors were more likely to survive to reproduce than those who didn't. These advantageous traits would then be passed on to offspring, conferring the same survival advantage.

One survival hypothesis is that the arts facilitate the acquisition of knowledge. Storytelling, the basis of literature, exercises the imagination, fosters useful thought experiments and produces certain kinds of social understanding (Dutton 2005, 2010; Pinker 2009; J. Carroll 2011). If so, then the individuals who engaged in storytelling behaviors might act more effectively and be more likely to survive, through their better understanding of situations encountered and of the outcomes of different courses of action. Another survival hypothesis is that art serves as a mechanism of social cohesion and cooperation. When we sing and dance together, for instance, we enhance our abilities and tendencies to cooperate. This gives us an advantage in survival (Dissanayake 1995, 120–125; McNeill 1995; N. Carroll 2014). After all, many of the activities important for life, such as hunting and gathering, defense from predators and battle with competing groups, are more effective to the degree they are done with the cooperation of other individuals. A third survival hypothesis, offered by Ellen Dissanayake, is that art behaviors, through a process of "making special," result in the better performance of social rites and creation of artifacts (Dissanayake 1990). Those individuals who cared enough to make things "special" (hand axes, for instance) would have made them better, and that conferred a survival advantage.

Perhaps the aesthetic preferences that function in our art-making might also be explained in terms of survival selection. Denis Dutton has argued that human aesthetic preferences for the habitats that are most conducive for life – those providing safety, water and nourishment – conferred a survival advantage. An aesthetic taste and preference for such a habitat would be expected to result in an increased tendency to search out and occupy the best and most hospitable habitats over more harsh and difficult environments, with their associated challenges for life. These aesthetic preferences might then be manifested in our landscape painting and photography. Dutton begins his book *The Art Instinct* with a discussion of a tongue-in-cheek survey by Russian artists Alexander Melamid and Vitaly Komar intended to discover what characterizes the most preferred paintings of each country. What they seemed to find was a nearly universal, worldwide preference for the same general type of pictorial representation: "a landscape with trees and open areas, water, human figures, and animals" (Dutton 2010, 13–15). This is precisely the type of landscape, according to Dutton, that is best for human survival.

Some behaviors and traits may be present not because of an advantage they conferred in survival, but one conferred in reproduction. In sexually reproducing species, for an individual's genetically based traits to be passed on to offspring, that individual must not only survive to reproduce, but must actually reproduce. That means finding a mate. Sexual selection was part of Darwin's thinking from the beginning of his evolutionary musings. In part, this was due to the difficulty of explaining sexual dimorphism and ornamental traits in terms of a survival advantage. How could natural *survival* selection explain those cases where the male and female had the same habits of life, but very different morphologies? If male and female gorillas, for instance, had the same selection pressures to survive, why was the male gorilla so much larger? Shouldn't natural selection have made them the same size? Neither were ornamental traits, such as the bright plumage and extravagant tails of birds, easily explained in terms of survival advantage. Surely the large, extravagant tails of peacocks and argus pheasants are a disadvantage in survival. Birds with these tails would be more vulnerable to predators, and have restricted mobility. We should expect that natural selection would eliminate these traits! For Darwin, the answer was that these ornaments, along with the sexually dimorphic traits, were instead formed and preserved through sexual selection (Darwin 1871; see also Richards 2012).

Darwin distinguished two types of sexual selection. First there is an *intra-sexual* selection by male combat, where the males fight for the right to reproduce with females. This form of sexual selection would be expected to produce sexually dimorphic traits such as the greater relative size of male gorillas, and the specialized weapons of combat such as the horns of elk. Second is an *inter*sexual selection, where males compete for the attention and favor of females (Darwin 1871, vol. I, 253–260). The males that are best able to attract females are most likely to reproduce. The male traits that are responsible for attracting females would then be passed on to offspring. This, according to Darwin, produced the feather and tail ornamentation we see in birds, as well as their elaborate songs. The preferences of the female peahens were in this way responsible for the development of the extravagant peacock tails. Perhaps the preferences of female songbirds were also responsible for the development and persistence of elaborate bird songs. For Darwin, who saw a continuity between humans and nonhumans, this was more than suggestive. In his *Descent of Man*, he explicitly considered the possibility that humans had undergone sexual selection and that that process was behind some of the most distinctive human traits, including art-related behaviors (Darwin 1871, vol. II).

Geoffrey Miller, in his book *The Mating Mind*, explores this possibility, arguing that human art behaviors are ornamental displays that reveal fitness. Only big-brained, healthy individuals can sing, dance, paint and compose poetry well. And the fittest will do these things the best. So a mating preference for individuals who are the best singers, dancers, painters and poets will in effect be a mating preference for the fittest individuals. And the offspring of a mating with those best at these arts will not just be fittest, but will also have the inherited mating preferences for those with artistic abilities. This produces a runaway process whereby each generation has a mating preference for art-making abilities and the associated high fitness. So, for example, we tend to enjoy singing and we find singers sexually attractive, which gives the best singers high social status and more mating opportunities, ensuring that both singing ability and the preference for singing ability will be represented in future generations. In this way, according to Miller, art is a kind of fitness display (Miller 2001).

A third evolutionary approach asserts that many of our art-related behaviors are not themselves adaptations for survival or reproduction, but are instead *by-products* of adaptations. While Steven Pinker allows that some art-related behaviors might be survival adaptations, making fiction for instance, he argues that other art-related behaviors are instead by-products. In some cases they are mere pleasure technologies, analogous to our taste for cheesecake, which is not itself adaptive, but is instead a by-product of our adaptive preference for fat and sugar. A taste for foods containing fat and sugar would have been advantageous for our ancestors in that it would lead to greater energy resources for those individuals with this taste preference, and would therefore be favored by survival selection. Cheesecake merely triggers the pleasure systems that moti-vated us to seek out sweet and fatty foods in the past. Similarly, according to

Pinker, our taste for music is a by-product of our language capacities and preferences, sensitivity to sound contours and emotional calls, auditory scene analysis and the pleasure circuits these systems engage. As Pinker puts it: "Music appears to be a pure pleasure technology, a cocktail of recreational drugs that we ingest through the ear to stimulate a mass of pleasure circuits at once" (Pinker 1997, 528).

Note a couple of things here. First, confirmation of these evolutionary theories is complicated and difficult. We would need to know about past environment and human behavior in those environments. We would need to know about the genetic basis for art behaviors. If there were no genes that inclined us to dance, sing, paint, etc., then these behaviors cannot be explained by these selection processes. Moreover, we would need to know the actual effects of the postulated adaptive behaviors on survival and reproduction. Did dancing and singing actually make us cooperate better in the past? And did storytelling really help us better understand and deal with social situations? Unfortunately, it isn't clear that we have all the required information. Second, while some of these accounts seem to conflict, whether music is a by-product of selection or a sexual adaptation, for instance, all three evolutionary approaches might have some application to art behavior in general. If so, the evolutionary story will be complicated. We return to that possibility in the next section. But for now we turn to the second way science has been used to understand art experience and behaviors, through psychology and neurobiology.

2.3 The Psychology and Neurobiology of Art

Psychology has long played a role in the philosophical thinking about art. Plato, for instance, seemed to rely on assumptions about human psychology in his critical discussion of mimesis in the *Republic*. In his *Poetics*, Aristotle seemed to assume a suite of stable intellectual, imaginative and emotional traits that have implications for what features drama would have. Hume seemed to assume a mostly uniform human nature, albeit with cultural variations, that serves as the basis for the regularities in aesthetic tastes. And Kant followed Hume in assuming a common human psychological nature in his claim that the experience of art involved the disinterested free play of the cognitive faculties. These assumptions about human psychology can be contrasted with the "historicist" assumptions of the 20th century, that there is no stable universal human psychology, and that we are "blank slates" to be written on by culture (Dutton 2005, 693–695). On this last view, and in contrast to the assumptions of Plato, Aristotle, Hume and Kant, a psychological study can reveal little or nothing about art in general, because art is "local" – specific to each culture and time.

If so, each culture, at any particular time, would be expected to have its own distinctive and idiosyncratic psychology of art. We should therefore not expect universal human aesthetic preferences.

Modern psychological approaches are more in the spirit of those who assumed a common and stable human psychology. Typically, their origin is identified with the work of Gustav Fechner and his 1896 book *Vorschule der Aesthetic*, with its description of experiments to test the "golden section" hypothesis, that proportions of 0.618 to 1 play a special role in human perceptions of beauty. Fechner tested this hypothesis empirically by asking subjects for their preferences among rectangles of various proportions (Vartanian 2014, 10). Fechner's experiments are examples of the first of three now common projects: the determination of aesthetic preferences, the identification of factors that operate in the formation of aesthetic judgments, and the discovery of the psychological and neurobiological mechanisms that operate in the creation and experience of art. Psychologists typically use three sources of information in service of these projects: phenomenology or felt experience, observational or experimental studies, and neurobiological investigations into mechanisms. While a comprehensive survey of all the modern psychological studies is not possible here, a few examples will be instructive.

In a series of experiments relative to the first project, the determination of aesthetic preferences, subjects were asked for their preferences relative to different versions of geometric grids, based on Piet Mondrian's paintings with their solid colors and strong horizontal and vertical lines. Using variations of the grids found in his paintings, researchers seem to have discovered that viewers prefer horizontal and vertical lines over oblique lines, and they prefer the spacing of the grids in Mondrian's actual paintings over the experimental variations (Locher 2014, 222–224). Other studies have been conducted to determine preferences relative to pictorial complexity and compositional balance, preferences relative to melodic originality and contour in music and the differences between the preferences of novices and experts (Kozbelt and Kaufmann 2014). In each of these cases, the researchers have assumed some sort of uniformity of aesthetic preferences. As we see in Section 4, the advocates of a program known as "neuroaesthetics" have taken this to an extreme in their attempts to uncover the universal laws governing aesthetic preferences.

The second project is the identification of factors that operate in the formation of aesthetic judgments. We might think that aesthetic judgments are straightforward reports of aesthetic preferences. But empirical evidence suggests this is not the case. Studies seem to show that exposure effects bias assessments. In one study, subjects were briefly exposed to some Impressionist paintings, and they judged the paintings they were exposed to as better than other paintings, but *only if* they were unaware of the exposure. Mere exposure seems to affect aesthetic judgments (Kieran 2011, 35; Lopes 2014, 25–26). Studies have also shown that reasoning about an aesthetic judgment seems to affect that judgment, making it correspond less well with long-term preferences. Researchers conclude that the reasons given were not explanations for actual preferences, but instead seem to serve social purposes related to social conformity and status. They are what Matthew Kieran called "snobbish judgments" (Kieran 2011, 36–37).

The third psychological project is the identification of the neurobiological mechanisms that operate in the creation and experience of art. Some recent research into the so-called mirror neuron systems, for instance, attempts to understand the mechanisms that operate in the observation of dance. Mirror neuron systems are specialized neurological circuits, discovered in other primates, that are activated both when an action is undertaken (such as reaching for a peanut) and when it is observed in others. Humans seem to have these same mirror systems that operate when engaging in an action and observing an action, as well as in having an emotion and observing an emotion (Winerman 2005; Clay and Iacoboni 2014, 316).

According to one experiment with ballet dancers and capoeira practitioners, the ballet dancers had more activation in their mirror systems while observing ballet actions than when observing capoeira actions. Conversely, the capoeira practitioners had more activation in their mirror systems while observing capoeira actions than when observing ballet actions. This would be expected because the ballet dancers had learned the ballet actions – training their mirror systems to ballet actions, but not the capoeira actions – while the capoeira practitioners had learned the capoeira actions – training their mirror systems to capoeira, but not the ballet actions (Calvo-Merino et al. 2005). In another experiment, female ballet dancers had more activation in their mirror systems when observing typical female ballet actions than when observing male actions, and male ballet dancers had more activation in their mirror systems when observing typical male actions than when observing female actions. Females and males had equal activation in observing actions typically learned by both female and male dancers (Calvo-Merino et al. 2006).

On the basis of these studies, Barbara Montero has argued that trained dancers may have an advantage in the observation of dance over nondancers, because the dancers have learned these movements and would therefore likely have more activation in their mirror systems. They can draw on their own learned proprioception to indirectly proprioceive the observed movements in a kind of kinesthetic empathy. This suggests, according to Montero, that learning the observed actions "can improve the aesthetic appreciation of beauty, grace, power and other such qualities." If so, perhaps dance training might make a critic better relative to the appreciation of the aesthetic properties of dance (Montero 2013, 170–174).

2.4 Criticisms of Scientific Naturalism

As we see in subsequent sections, there is much more to the biology of art. But these two approaches, based on evolutionary thinking and psychology broadly construed, have gathered the most attention from philosophers and nonphilosophers alike. From some philosophers there has been criticism. Some worry whether the science is settled or correct (D. Davies 2013, 200–201). Our evolutionary theories, for instance, are still largely speculative, so we cannot say for certain whether dance is a survival adaptation, a sexual adaptation or a by-product. And our understanding of mirror systems is still incomplete so we don't know precisely how they work. All this is undoubtedly true, but absolute certainty has never been a realistic goal in science, and the sciences have been fabulously successful despite their fallibility. Moreover, alternative approaches seem to have their own problems. A priori philosophical analysis, for instance, has had a mixed history. One need only read Descartes's *Meditations* (written, he claims, by shutting himself in a small room) and its proofs of God's existence to become aware of the shortcomings of that approach. Even those who agree with Descartes's conclusion are rarely happy with his analysis.

Suppose for the moment that the scientific accounts are correct and we have sufficient evidence to establish them. Surely the evolutionary approaches provide insight into the ubiquity of art-like behavior in humans. And surely the psychological studies and experiments reveal important things about our engagement with art, the basis of our aesthetic judgments and the mechanisms that underlie our experience of art. We could grant all that but still be dissatisfied with our scientific naturalism, and on two grounds. First, what can science tell us about how we *should* engage with and experience art? Second, what can science tell us about the conceptual basis of art? If a scientific naturalism cannot give answers to either question, then perhaps it cannot do the important philosophical work. Philosophically, a scientific naturalism would then be relatively trivial.

We might think of the first objection as the *normativity problem*: science may tell us about how we actually experience art, but it cannot tell us how we *should* conceive of and experience art, or what makes an artwork good or bad. According to George Dickie: "No matter how many data are collected, they still remain descriptions (the *is*) and no normative principles (the *ought*) can be derived from descriptions alone" (Dickie 1962, 295). (See also Dorsch 2014, 87.) Suppose evolutionary theory is correct in claiming that dance conferred a survival benefit by enhancing cooperation. Does that mean we *should* dance in ways that cooperation is enhanced? If so, then perhaps we shouldn't encourage or allow solo dances. Suppose music in general and singing in particular are fitness indicators that function to attract mates. Does this imply that we *should* sing to show our fitness so that we get additional mating opportunities? Suppose instead that music is a mere pleasure technology. Does that imply we *should* make and listen to music? Or does it imply that we should resist these hedonist tendencies, and we *shouldn't* make and listen to music? Does it also imply that some music is *better* than other music – perhaps through greater stimulation of pleasure circuits? The problem here is that in none of these cases do the normative conclusions seem to follow automatically from the empirical data. It is an open question whether we should dance to enhance cooperation, or sing to maximize mating opportunities.

We can ask the same questions about the psychological studies and experiments. Research may tell us that viewers tend to prefer the actual Mondrian geometric grids over the experimental variations. Does that mean that we *should* prefer the actual Mondrian grids? Suppose the research is correct that our mirror systems affect our experience of dance. Does that mean that we *should* take dance classes to better enjoy *Swan Lake*? On the other hand, maybe this kinesthetic empathy through mirroring activation is distracting (D. Davies 2014, 73). Perhaps we should instead focus on the formal qualities of the dance. Suppose we find out that we prefer paintings that we have had unconscious exposure to. *Should* we prefer these paintings? Or does this exposure produce an irrelevant bias? In all these cases it isn't obvious how the normative conclusions follow automatically from the facts. We return to this problem in Sections 5 and 6.

The second philosophical objection is that to do art we need a concept of art, and that is not something that can be studied by science. According to Graham McFee, to make a dance, or to see something as dance, requires the possession of the *concept* of dance (McFee 2011, 14). Mere rhythmic movement is not dance. Dance is a kind of action with rules, and one must act in accordance with those rules to be dancing. If so, then we may learn all there is to know about the psychology of making and observing dance, and we may have the right evolutionary account of dance, but we would still know nothing about the conceptual preconditions of making and experiencing dance. The evolutionary and psychological accounts of dance don't study art concepts and how they are used. They study behaviors, survival, reproduction and neurobiology. McFee concludes that the sciences have nothing of real philosophical importance here.

McFee is surely right in that a brain image shows various kinds of neurological activity, but it does not and cannot directly show *conceptual* activity. We cannot look at the brain images and see concepts in play. Similarly, the evolutionary accounts do not seem to incorporate concepts along with genes, survival and reproduction. If we want to know how art concepts work, we need to look elsewhere. But the conceptual objection also relies on some assumptions we may ultimately question. First is the assumption that to dance one must have a concept of *dance*. This raises obvious questions. What are *concepts* and how do we act on their basis? And must we really have a concept of dance to dance? If so, then in order to determine if the famous Masai jumping dance is really dance, we would need to find out about the Masai concepts, and see if any of them count as a *dance* concept. These questions engage issues that go beyond this section, but we return to some of them in Section 5.

2.5 Conclusion

Neither of these philosophical objections to recent scientific approaches to understanding art implies that the science isn't worth doing. What they seem to imply is that the sciences cited cannot do everything. In some sense this is uncontroversial. We don't demand, for instance, that physics tell us everything about why peacocks flash their large, extravagant tails at peahens – even though at some level physics is most certainly involved. A charitable reading of these criticisms is that they were directed at the science enthusiasts who naïvely thought the science could replace the philosophy. While I don't believe the science can replace the philosophy, I am among those who think the science can do more than these critics allow, even if the evolutionary and psychological theories criticized here seem to have the limits the critics claim. As we see in what follows, the scientific contributions to the normative and conceptual understanding of the arts will come from other parts of the biological framework – the ecological in particular. But before we turn to the ecology of art, it will be useful to take a closer look at the evolutionary theories of art in the next section, and psychological and neurobiological accounts in Section 4.

3 The Evolutionary Framework
3.1 Introduction

Because the naturalism proposed here starts with our best scientific theories, and evolutionary theory is one of those theories, we revisit some of the evolutionary approaches introduced in Section 2 that explain art in terms of survival or sexual selection, or as a by-product of adaptation. These selection-based accounts often seem to assume that some art behavior evolved to serve some *single* function relative to survival or reproduction. But a multifunction approach, where art serves multiple evolutionary functions, may be better. In this section we see how that might work. Then we sketch out an expanded evolutionary framework that includes genetics, epigenetics, development and ecology.

Why adopt an evolutionary approach to art at all? A full vindication will ultimately depend on its explanatory success, but there are some reasons to at least consider an evolutionary framework. First, and most obviously, is the apparent universality of art behaviors across cultures (Dutton 2010, 29–30). Dancing, singing, storytelling and decorating, for instance, seem to be found in cultures worldwide and in broadly similar forms. It is easy to see the rhythmic movement and leaping of the Dinka in the South Sudan as dancing, and in the same way we see Hungarian folk dancing, Irish step dancing, classical ballet and hip-hop all as dancing. The vocalizations we find in cultures all over the world are similarly easy to identify as singing, albeit with different technical and stylistic features (Jordania 2011). The decorations and sculptures of cultures worldwide similarly look like the same kinds of things. And while there are clearly differences among these behaviors, and they may have different meanings and significance in different cultures, *that* all people seem to dance, sing and tell stories surely cannot be *just* a product of culture. Perhaps art is like language in that all human cultures have language, even though there are many variations in terms of its expression – sound patterns, structure, vocabulary and meaning. If we turn to evolution to understand human language, why not turn to evolution to understand human art?

A second reason an evolutionary approach is at least plausible is the spontaneous development of some art behaviors in children. Music-related abilities such as sensitivity to pitch and prosody seem to develop before full-blown language abilities (Trehub 2003). And almost all of us have the ability to hear sound and singing as music, and do so from a very young age. The inability to do so, "amusia," is striking and rare (Sacks 2007, 98–119). Similarly, children seem to spontaneously and skillfully engage in make-believe in ways that are similar to what we might see in a stage play (Dutton 2010, 106–108).

A third reason to consider an evolutionary approach to art is the pleasure we get from it. We spend so much time playing or listening to music, watching drama and dancing, in part because of the pleasure it brings. Since pleasure is an evolved mechanism for motivating behavior, perhaps it is a sign of some evolutionary function of art. After all, those who found pleasure in eating were more likely to survive than those who didn't, and those who found pleasure in sex were more likely to reproduce than those who didn't. *Perhaps* those who found pleasure in art were more likely to succeed at something, or to function better in some way, than those who didn't.

3.2 Art in Other Species

But before we return to the evolutionary theories of *human* art behavior, we might ask if other species exhibit art or at least art-like behaviors. If other species engage in the arts, then we have an additional reason to consider an evolutionary approach. There are obvious candidates. Birds seem to sing, and in ways that involve cultural learning and variation (Prum 2013, 823). Gibbons and whales vocalize in ways that sound like song. Australian bowerbirds build and decorate bowers in what seems like an artistic way, using color themes and formal design. Sage grouse and peacock spiders seem to do something like dance in their stylized and rhythmic movements.

But are these *really* art? The proposed definitions of art we considered in Section 1 don't directly address this question about art in other species, but we can ask if the postulated functions of art – as representation, expression and production of pleasure or aesthetic experience – are unique to humans. Do bowerbirds intend to represent or express anything with their bowers? Do peacock spiders dance as expression? Do songbirds sing to produce aesthetic experience? This seems implausible. Moreover, aren't these nonhuman behaviors formulaic and inflexible, while human art behaviors are creative and flexible? (S. Davies 2012, 10–15, 31–34). So even if these functions don't successfully define art, there may be too much difference between bird songs and human songs, between peacock spider dances and human dances and between bowerbird decorations and human decorations to treat these all as the same kind of thing – as *art*.

There are some obvious problems with these attempts to distinguish these animal behaviors from the human. First they just assume that bowerbirds and peacock spiders don't have the right kind of intentions or aesthetic emotions, and don't have the right kind of creative flexibility. One easy way to do that is to just assume that *human* intentions, aesthetic emotions and creative flexibility are the standards for art. But this is begging the question. Of course, only humans have human intentions and human aesthetic experience. Moreover, there are good reasons to believe that some of the art-like behaviors are not as different as assumed. The dances of the sage grouse and peacock spiders, for instance, cannot be formulaic in the sense that they are all the same, or they wouldn't function as they do in sexual selection, to distinguish the individuals displaying the movement. Why not think of the variations in these dances as involving creativity? Second, as we saw in Section 1, it has not yet been established that intentions to express, educate and produce aesthetic experience are necessary even for *human* art. If so, it is surely too much to require these things for nonhuman art.

It is more clear that the procedural definitions of art cannot be applied to nonhuman art-like behaviors. It is implausible that songbirds, bowerbirds, gibbons and peacock spiders have art traditions or institutions that they could be working within to make what they do count as *art*. But again, it has not been established that a necessary condition for art is found in the procedures established by art traditions and institutions. Recall that we might plausibly regard the first human cave painting as art, even though there could have been no preexisting traditions or institutions to make it art.

Finally, we might worry that if having the concept of *art* (or of the individual arts) is necessary to do art, then since songbirds don't obviously have a concept of song (or art), and peacock spiders don't obviously have a concept of dance (or art), they are therefore not singing, dancing or doing art in general. But this worry raises the question: what does it mean to have a concept of *art*? To answer this question we need to say what concepts are, and what it means to have one. On one standard account, concepts are abstract things associated with propositional content. Having a concept is equivalent to knowing the content of that concept (Margolis and Laurence 2014). Having a concept of *water* would then be knowing that it is H_2O. If this is right, then to have a concept of art would require having something like a definition of art. The failure to find a satisfactory definition of human art, though, suggests that *we* don't have a concept of art in this sense either. If so, how can we require nonhuman species to have a concept of art in this sense?

But other ways of thinking about concepts may be more useful here. Perhaps to have a concept is simply to have an ability or capacity (Margolis and Laurence 2014). If so, nonhuman creatures can have concepts. A dog, for instance, would have the concept of *dog* if it can distinguish dogs from other things. A bird would have the concept of *singing* if it can distinguish singing from other behaviors, and a peacock spider would have the concept of *dancing* if it could distinguish dancing from other behaviors. A cave painter would have the concept of *painting* if he or she could distinguish *cave painting* from other activities. We cannot adjudicate the different theories of concepts here, but it is important to see that there are ways of thinking about concepts that allow for nonhuman animals to have the relevant concepts in ways that factor in their behaviors. This would allow for all sorts of creatures to engage in art as long as they can make the right kind of distinctions. We might go even further, as Richard Prum does, when he argues that art is ubiquitous across biodiversity in biotic display, where "art consists of a form of communication that has coevolved with its evaluation" (Prum 2013, 818). According to Prum, we can see this "biotic art" in all the courtship displays we see, and even in flowers, which have evolved through the shape and color of their petals to attract pollinators by visually stimulating their sensory and cognitive systems in particular ways. (We return to this idea in the last section.)

3.3 The Evolution of Human Art

A full evolutionary account of art requires that we know the circumstances of its development, including *when* it evolved. There seems to be some consensus that humans began to engage in many art-like behaviors at least by the beginning of the upper Paleolithic, somewhere around 35,000–40,000 years ago, producing cave paintings, rock art, decorated weaponry, carved bone and stone sculptures. But some candidates for art appeared much earlier, including some things that *might be* carved figurines from more than 300,000 years ago – the "Venus of Tan Tan" – and from more than 250,000 years ago – the "Venus of Berekhat Ram." Decorative and perforated shells seem to have appeared more than 75,000 years ago. Marked ostrich eggs with geometric patterns seem to be at least 60,000 years old (Puta 2015, 48–52). There is, of course, no record of human literature until written language appeared just a few thousand years ago, so we cannot know when storytelling first appeared. And since the beginnings of song and dance left no traces, their origins are a matter of speculation. Joseph Jordania, though, argues that singing first appeared millions of years ago, in our primate tree-living ancestors, and humans, alone among the ground dwelling primates, retain their singing (Jordania 2011, 89–91). But we do not and cannot have direct evidence of the first singing, or of the first dancing and storytelling.

It is not obvious that there is a single origin of some human behavior we might describe as art. But what we know doesn't rule one out either. The problem is that there is simply too little information to reconstruct with confidence the origin of art or art-like behavior in the human branch of the evolutionary tree. These problems, though, are not unique to the evolutionary reconstruction of art behavior; they plague *all* attempts to reconstruct the evolutionary past. We don't have direct traces, for instance, of human evolution. We simply have skeletal remains of human-like creatures found in various places, skeletal remains that are similar in some ways and different in others. From this, we *infer* human origins. This inference, like the inference of art origins, is neither simple nor easy. A quick survey of how our human origin theories have changed will make this clear.

So how should we explain the origin of art-like behavior in humans? In Section 2 we briefly discussed three main evolutionary approaches to art, based on the ideas that art has survival or reproductive value, or is a by-product of something that had survival or reproductive value. That discussion barely scratched the surface of evolutionary thinking about the arts. Here we adopt a different strategy. Instead of grouping evolutionary theories by postulated mechanism – survival selection, sexual selection or by-product of selection – we start with clusters of related art behaviors. This strategy makes it easier to see how art-related behaviors may serve multiple evolutionary functions. So in this section we look at evolutionary theories of particular art behaviors, primarily storytelling and music and dance, with a gesture toward the visual arts. In each case, I try to sketch out a plausible, comprehensive evolutionary account of the art behavior complex – the behavior itself, its underlying capacities and the tastes or preferences that guide it.

3.4 Storytelling

We could begin our analysis here with highly specific behaviors. In the literary arts, that might be writing and reading poems, histories, religious texts, novels, plays or films. That starting point is clearly problematic, though. These specific behaviors depend on the environment. We could only write a poem, novel or play after writing was invented and only in those cultures that had these traditions. If we want to understand this cluster of behaviors, perhaps we can best start by thinking about them in more general terms – as instances of storytelling. This is the art behavior most widely regarded as an evolutionary adaptation. Even Steven Pinker takes it to be an adaptation, though he generally thinks of the arts as by-products of adaptation (Pinker 1997, 541–543).

One possible evolutionary function of storytelling is that it helps us to acquire the knowledge needed to navigate a complex and risky world. First, we could learn from the content of stories about risky encounters we might have with nature, the conflicts we might have with other people or difficulties we might encounter in maintaining relationships with family, mates and friends. Second, we could also learn the social roles we would be expected to play – being a parent, a mate, a rival, a leader and so on – and how to fulfill those roles. Third, we could learn from the counterfactual thinking in fictions the various strategies in dealing with problems. What options do we have in dealing with a romantic rival, a jealous lover, an angry child, and so on? (Pinker 1997; Dutton 2010; J. Carroll 2011). Perhaps the person who has encountered stories that prepare him or her for a complicated and risk-filled life would have an advantage in navigating this world, in survival and reproduction, over the person who had not been so prepared.

One initial problem with this adaptation hypothesis is that it doesn't seem to recognize that these advantages conferred in storytelling are for the person who *hears* the stories, but not for the person who *tells* the stories. If storytelling behavior benefits the hearer but not the teller, how could it evolve? We would expect that many people would want to hear stories but very few would be anxious to tell them. Insofar as there is competition for resources or mates, there might even be an advantage to keeping useful information for oneself. If so, it is unclear how there would be stories to hear, whatever the advantage in hearing them!

There are two ways to get around this problem. First, a parent might benefit genetically through the future success of their genes, by telling these stories to their children, who, based on the content of the stories, would then be better prepared to navigate the complicated and risky world and have their own children. These children would have an advantage over those children who did not hear stories from parents. The children of storytellers might also benefit from other psychological and developmental functions. Perhaps listening to stories facilitates cognitive and emotional development and hones the imagination, which then has general survival benefits for both the children and the parents' genes (Robinson 2005; N. Carroll 2014; Carroll, Moore and Seeley 2014). One might worry, though, that not all stories are told by parents to their children. How can we explain this storytelling to other people in general? One answer might be that offspring of parents who tell stories indiscriminately would still have an advantage over the offspring of parents who don't tell stories indiscriminately. The former would likely hear more stories simply because they are more often around a storyteller. Second, there might also be an advantage in telling stories that represent – or misrepresent – the world in particular ways. Stories that represent the teller as having high status, as being particularly brave or smart or as possessing other unusual skills might generate an advantage in interactions with other people. They might enhance the storyteller's social status, motivating others to cooperate with them. If so, this kind of storytelling might present an advantage in survival. An elevated status might also be an advantage in reproduction, a special kind of cooperation.

As we saw in Section 2, a trait could be explained, not by an advantage conferred in survival, but by an advantage in attracting mates. Some traits are fitness indicators in that only individuals of high fitness can produce them. Assuming fitness and preferences are heritable, a mating preference for fitness indicators will produce offspring that are more fit, and with a preference for the fitness indicator. If storytelling were a fitness indicator, in that it depended in some way on the fitness of the individual telling the story, then it could generate the runaway process described in Section 2. There is a plausible case to be made here. Surely storytelling abilities are dependent in part on the size of vocabulary, the ability to put words together according to grammatical rules and the ability to understand social roles and strategies in dealing with social problems. And surely these abilities are dependent on having a large, healthy brain. But large, healthy brains are very costly, requiring something like 15% of oxygen intake, 25% of total metabolic energy and 40% of blood glucose – even though they only comprise about 2% of our body weight (Miller 2001, 134). Moreover, many things can interfere with brain development, including parasites and pathogens. So only those individuals who can acquire sufficient nutrition and who are resistant to parasites and pathogens can grow a good brain – and tell good stories.

The preferences we have in the content of our storytelling might also have an evolutionary explanation. We would plausibly find most interesting those stories about things that matter most to us – survival and reproduction. And with humans in their social contexts that means stories about romance, children, parents, rivalries and so on. This is one of the commitments of the approach to interpretation known as "literary Darwinism" and advocated by Brian Boyd and Joseph Carroll (Boyd 2005 and J. Carroll 2011).

There are several things to notice here. First, what is important in this account is that the relevant traits confer *evolutionary* advantages if they provide advantages in survival and reproduction. But storytelling may also perform other beneficial functions, such as giving pleasure and helping us understand other times and places. Those functions are only relevant to the evolutionary account insofar as they serve purposes relative to survival and reproduction, though. Second, sometimes storytelling might actually be dysfunctional, if we hear and believe stories that misrepresent the world in certain harmful ways (S. Davies 2012, 170). But notice an evolutionary adaptation need not *always* serve the function that drove its development. What is required is that at some time and *on the whole*, storytelling presented an advantage in survival, reproduction or both. Third, we can think about how each component of storytelling might have developed. The mere *capacity* to tell stories would not have been favored by selection processes unless it conferred an advantage through its expression in behavior. Preferences or tastes can also be given a selection-based explanation. Those individuals who prefer to listen to certain kinds of stories might benefit more from the storytelling than those who have different tastes and preferences. If so, those preferences may spread throughout the population. Similarly, tastes and preferences for stories that are more difficult to tell might be favored by sexual selection, if they function better as fitness indicators about their tellers.

Notice that storytelling is also a by-product of selection in that it is dependent on faculties that evolved on the basis of other functions. Most obviously it depends on the basic ability to represent people and events in language. More basically, it also relies on the ability to break up a sound stream into words, the capacity to attribute a mind and mental states to other people and more. So a by-product explanation is not necessarily incompatible with an adaptation explanation. The correct explanation of storytelling might be complex and based on multiple functions. It is possible, for instance, that storytelling began as a by-product of language skills and higher-level cognitive functions, that then became the basis for fitness display and was favored by sexual selection, after which it acquired other purposes, the education of offspring, the manipulation of listeners and the establishment of social status that might serve survival purposes. If so, we cannot explain storytelling simply as either a survival or reproductive adaptation. It may be both of these things, and in multiple ways, as well as a by-product of other adaptations. If so, perhaps there is a similar, multifunctional explanatory pattern when we look at other art behaviors.

3.5 Dance and Music

Evolutionary accounts of dance usually begin with the identification of social functions. Dance, along with music, is often claimed to enable or enhance cooperation (Dissanayake 1990; McNeill 1995; Mithen 2006; N. Carroll 2014). Those individuals who danced, made music or sang together were generally better able to cooperate with each other in other activities. So the individuals with the greatest capacity to do these things would have an advantage in doing things that were important and that require cooperation – hunting, gathering, fending off dangerous animals, fighting battles and so on. In this way survival selection favors the capacity to dance and to make music together.

But the correlation of singing, dancing and music with cooperation might happen several ways. Perhaps those individuals who had an antecedent ability to cooperate may have had an advantage, and having that ability also made them inclined to dance and sing together. In this case, the dancing and singing behavior would simply be a manifestation of the more general, functional ability to cooperate. It wasn't the dancing and singing itself that conferred an advantage in survival. It was the underlying capacity to cooperate. Or alternatively, perhaps the dancing and singing itself conferred the advantage. Those individuals who had the capacity to dance and sing together – and actually did so – were able to better cooperate *as a consequence* of their singing and dancing behavior. They would then have an advantage over those who merely had the capacity to cooperate, but did not engage in the singing and dancing behavior.

Those individuals who dance and sing, and teach their children to dance and sing, would then have more and better cooperative relations that give their genes an advantage over generations. If so, we might think of dance and song then as an evolved mechanism for the promotion of cooperation.

Dance and music might also function at another level in making possible the construction of social groups. According to Noël Carroll (2014), a society's functioning depends on the ability of its members to make converging judgments and engage in coordinated action. These are facilitated by emotion. Fear, for instance, motivates judgments of risk and harm, and stimulates action. The arts in general educate emotions, according to Carroll, but some arts, music and dance in particular, generate converging emotion, what he calls "emotional contagion." When we sing and dance together we come to feel the same emotion, and this makes us respond *as a group*. Carroll claims this behavior should be conceived as a group-level adaptation, making some *groups* fitter than others. Those groups that dance and sing together function better as a group, consequently becoming more able to survive *as a group*. This turn to group selection involves many complications that cannot be addressed here (see Sober and Wilson 1998), but for now, it may be sufficient just to note that the behavior of dancing and singing might make possible, not just enhanced cooperation but also the formation and maintenance of discrete social groups. If so, then it is not surprising that we march together in military groups, sing together and play music within our fan groups at sporting events, and within our religious groups in church.

Dancing and singing might also be a product of sexual selection. If it is difficult to dance, sing and play music well, and if that requires a healthy, well-functioning brain, then a mating preference for good singing and dancing will also be a preference for high fitness. A runaway process might then be generated where dancing, singing and music-making abilities evolved along with preferences for the associated behaviors. Even so, it is clear that dancing and singing are also dependent on other antecedent adaptations, such as bipedalism, beat entrainment, vocalization skills and the ability to parse a sound stream, and are therefore by-products of other adaptations.

Like storytelling, the evolutionary account of song, music and dance might be complex. Because these behaviors are dependent on other adaptations, they are in some sense by-products. But perhaps each behavior also had multiple evolutionary functions at any single time and over different times. Dance, for instance, may have been a side effect of bipedalism in humans, but it then could have come to be favored by sexual selection, based on its value as a fitness indicator. This is plausible given that dance-like behaviors in other species, sage grouse and peacock spiders for instance, are examples of sexual display. But then dance could later serve to enhance cooperation, with survival benefits at both the individual and group levels. Similarly, song may have started out as a by-product of expressive vocalization, but later came to be subject to sexual selection as a fitness indicator. Then it may have been used to manipulate the emotions of others and to foster cooperation (Mithen 2006). And in this case, as with storytelling, there may also be benefits that don't have evolutionary functions, such as the ubiquitous manipulation of our own emotions and moods by listening to music on our electronic devices. Music and dance may be pleasure technologies as well.

While there isn't space to consider it here, there may be a similar multi-function evolutionary account of the visual arts. The capacities to paint and draw are obviously by-products of traits that evolved for other reasons, the grasping capacity of a hand with an opposable thumb used for climbing, for instance. And the representation of faces in painted portraits, for example, surely depends on our abilities to recognize faces, which has obvious value in social circumstances. The actual behaviors *may* have performed evolutionary functions relative to sexual or survival selection, and the manifested tastes or preferences manifested *may* also have evolutionary explanations. And perhaps, as we considered in Section 2, our landscape painting preferences are products of the adaptive value of habitat preferences. If so, we might also give a plausible multifunction evolutionary account of painting, sculpture and the decorative arts.

3.6 The Full Evolutionary Framework

These evolutionary accounts just sketched out are at least plausible, but they are not fully satisfying for two reasons. The first is that they are speculative in that we simply don't know enough about the circumstances of the evolution of art-like behaviors to be certain about the evolutionary functions they served. But second, these accounts also make other assumptions that we cannot yet confirm. For an evolutionary account based on survival or sexual selection to be true, there must be some genetic basis for the specified behaviors. Is there a dancing gene, for instance? One study from 2005 found genetic differences between dancers and nondancers that affect sociality and spirituality (Bachner-Melman et al. 2005). While these genes may be important to dance, they are hardly dancing genes, and it would be absurd to propose that a single gene is responsible for all the capacities that dancing requires. An evolutionary account requires a suite of genes that makes a difference in the behavior of organisms, inclining those with these genes to some advantageous behavior that can be transmitted to offspring, and inclining their behaviors as well. But even if we knew the genes responsible for the development of an art behavior, genes can sometimes be expressed and sometimes suppressed. So we would need to know the epigenetic factors that influence development – the regulatory gene networks and environmental factors that govern the expression of genes. Unfortunately, too little is now known about these factors that influence the expression of genes.

The environment is important in other ways as well, in that it may guide how a particular genetically based behavior develops. An environment with a highly developed and codified set of dance practices and conventions (classical ballet perhaps) will produce a different set of dance behaviors than an environment that has only informal folk dance practices and conventions. Moreover, the operation of natural selection is always dependent on the environment as well. What may be an advantage in one environment may not be an advantage in another. The bottom line is that we need to look at the *ecology* of art, how individuals interact with each other in the art contexts of their environments. As we see in Section 5, what is distinctive about humans is how they have modified, or engineered the environments in which art behaviors develop.

An evolutionary approach based on selection processes and the genetic basis of behaviors has one implication that should be explicitly recognized. Modern evolutionary theory is widely recognized to be based on "population thinking" in that evolutionary change can occur in a population only if there is substantial genetic variability within that population. Some individuals in a population have a genotype that gives them an advantage over the others, and that genotype gets preserved in later generations. This fact implies that we should not assume that all humans have precisely the same capacities for art behavior, or the same tastes that guide this behavior. There might in fact be substantial variation relative to any trait. As we see in what follows, this fact is important when we look at the evaluation of art in Section 6.

3.7 Conclusion

So how should we respond to this multifunctional, evolutionary framework for the arts? An evolutionary skeptic might see it merely as a rationale for the many "just so" stories based on little evidence and much speculation. Perhaps Steven Pinker, even though he isn't an evolutionary skeptic in general, is right in his claim that most of the arts are mere pleasure technologies like cheesecake and pornography, invented to satisfy desires that have other evolutionary functions and origins (Pinker 1997, 528). The problem with Pinker's claim here, though, is that it isn't at all clear that art is biologically useless.

On the other hand an evolutionary enthusiast might see all these adaptation theories as simply more evidence of the power of evolution to explain all human behavior. But the fact that we can construct plausible adaptation hypotheses does not by itself imply that any of them are correct. The right attitude, I believe, is somewhere between evolutionary skepticism and enthusiasm. Unless we want to reject modern biology and its evolutionary framework, there must be some evolutionary story to tell. But that evolutionary story is not *just* one of adaptation by selection. The story will be a complex one, involving genetics, epigenetics, psychology and neurobiology and ecology. The next two sections continue to fill out this framework, with the psychology and neurobiology of art in Section 4, and the ecology of art in Section 5.

4 The Psychology and Neurobiology of Art

4.1 Introduction

In the previous section we briefly looked at some clusters of art behaviors and how they might have multiple evolutionary functions related to survival and reproduction, while still also being based on traits that evolved for other functions. These evolutionary accounts all implicitly assume genetically based mechanisms underlying the art behaviors. Otherwise there would be no inheritance of the advantageous art-related traits. Genes, of course, do not directly produce art behaviors, but instead code for the proteins that make the bodies and brains that engage in art behaviors. These bodies and brains then have particular psychological tendencies and neurological mechanisms that influence how art behaviors get expressed. In this section, we take a look at some prominent theories about these tendencies and mechanisms.

Before we do so, though, we need to briefly consider how genes influence development. Two extreme views should be rejected from the start. The first, often referred to by its critics as "genetic determinism," treats development and behavior as fully controlled by genes. The second, referred to by its critics as a "blank slate" approach, treats development and behavior as fully determined by environment. On the first view, at one extreme, behavior is to be explained by genes alone, and on the second view, at another extreme, behavior is to be explained by environmental influences alone. But biology tells us that development *always* involves *both* genes and environment. We can see this in the interactionist model of development of biologists Charles Lumsden and E. O. Wilson (1981).

According to this model, genes constrain development in that they produce a limited range of outcomes – a "norm of reaction" – in survivable environments. So for any trait there is a range of possible developmental outcomes that could occur, depending on the features of the environment in which development actually occurs. And within this reaction norm, genes may also bias development, making some outcomes more probable than others. Lumsden and Wilson tell us that we can think of the genetic constraints and biases in terms of *epigenetic rules* that influence development without fully determining it. There are the *primary epigenetic rules*: the genetic instructions for building the human retina, for instance, with its distinctive pattern of rod and cone photoreceptor cells (Lumsden and Wilson 1981, 36). These primary rules also are responsible for the general human reliance on vision over the senses of taste and smell for information about the environment. They give instructions for building a primarily visual brain rather than a chemo-sensitive brain.

The primary epigenetic rules are more automated and less sensitive to environment than the *secondary epigenetic rules*. We see the operation of these secondary rules in language development. Human genes make language learning possible within certain constraints, but *which* language is learned depends on the environment. And research seems to show that our abilities to hear particular phonemes depend on the environment – which language we are exposed to at the critical learning phase. For instance, at a certain age, babies can hear distinctive sounds of both Hindi and English, but as one language is learned, they lose the ability to hear the sounds associated with the language they did not learn (Werker and Tees 1984). It seems that babies are genetically biased to learn to hear and parse the sound stream of some language or other, but the actual perception of language is a product of the environmental stimuli. We can think of it this way: the presence of a language mechanism with certain general tendencies is due to genes, but how the mechanism actually develops and works depends on the environment.

In Section 2, we discussed some approaches to art that relied on empirical assumptions about human psychology, beginning with Plato and Aristotle, then through Hume and Kant, and on to the beginning of modern psychology in the work of Gustav Fechner, who studied preferences relative to the "golden section." We then considered three different projects in modern psychology and neurobiology: the determination of art preferences, the uncovering of factors that affect aesthetic judgment and the discovery of mechanisms that underlie the making and experiencing of art. In this section, we start by looking at the general contours of a recent research program known as "empirical aesthetics," and then turn to some of its most prominent general theories. We then look at some approaches that begin with what we think we know about the neurological mechanisms underlying art behaviors and aesthetic experience.

4.2 Empirical Aesthetics

Roughly, empirical aesthetics is the study of aesthetic experience using observation and experiment. But what is aesthetic experience? We could construe it narrowly as the perception of beauty and the experience of the sublime, found exclusively in humans (S. Davies 2012). On this construal aesthetic experience is rare and difficult, and typically occurs only in special places – museums, theatres and concert halls. In empirical aesthetics, though, aesthetic experience is usually understood more broadly as sensory experience with a hedonic response (Shimamura 2014, 4). If so, eating a snack, drinking cold water on a hot day and smelling flowers might all count as aesthetic experience, insofar as they involve pleasurable sensory experience. This construal of aesthetic experience is too broad for our purposes here, though. Perhaps we can simply recognize that when we think about art, we are sometimes interested in particular kinds of aesthetic experiences related to our engagement with art.

Because human perception is very much oriented to vision (rather than taste, touch or smell), much modern empirical aesthetics has focused on visual experience. Early on it adopted a "bottom-up" starting point, as in Fechner's study of the golden section that was based on preferences for rectangles of varying proportions. This general approach begins with elemental sensory stimuli, typically shapes (such as rectangles) or colors, then tests for preferences, and tries to understand these preferences. But there is also a "top-down" approach, associated with Gestalt psychology, that denies perception can be dissected into basic elements and that treats visual experience as driven by the active organization of perception on the basis of learned schemas or conceptual frameworks that establish expectations. When we go to a museum, for instance, we activate the appropriate museum schema, and that tells us what to look at and how to see the objects in the museum (Shimamura 2014, 14–18). And when we look at blotches of color in a painting, we may see them as objects or people, based on our conceptual framework and what we know about objects and people. Most recent research seems to have elements of both the bottom-up and top-down approaches. We see this in the general theories in empirical aesthetics that attempt to explain the perception, emotion, cognition and reward associated with art experience.

4.3 Arousal, Prototype and Fluency Theories

In the early 1970s, psychologist Daniel Berlyne argued that what is distinctive about art experience is the psychological arousal it produces, based on the "collative" properties of artworks – their novelty, complexity, surprisingness, uncertainty, conflict, unfamiliarity and incongruity. A painting, for instance, may represent something in a novel or surprising way, and this produces psychological arousal. If it is the right amount of arousal, it will result in pleasure or interest. In general, arousal is a product of a mismatch between incoming information – the perception of the art stimulus – and what was expected. Whatever is surprising, unfamiliar or incongruous generates arousal, which then produces pleasure or interest, and in turn guides behavior, perhaps just to keep looking or listening.

Berlyne tested his theory in a series of experiments on both untrained under-graduates and subjects with more art expertise, and discovered that the optimal amount of arousal for the former was generated by simpler artworks and for the latter by more complex artworks. He took this to imply that knowledge affects aesthetic experience. This is not surprising, given that arousal depends on expectations, and expectations depend on how one conceives the world and what one knows. Berlyne also distinguished those responses that found some-thing "pleasing" and generated pleasure from those responses that found some-thing "interesting," which affected attention. He regarded these as different types of responses to artworks and their collative properties, but lamented that he lacked a good explanation as to why a set of collative properties might result in "pleasingness" in one case and "interestingness" in another (Shimamura 2014, 16; Silvia 2014, 254–257).

A second general theory, developed in the 1980s, is based on the idea that we have a preference for prototypes – objects that are typical of their category (Martindale and Moore 1988). Evidence for this theory comes partly from research unrelated to art experience, which found that people generally prefer typical faces, colors, objects and animals over the atypical. Some research indicates this is true of artworks as well. In one study, subjects preferred typical surrealist paintings over the atypical, even when all the paintings were new to the subjects (Silvia 2014, 257–258). And it seems that subjects prefer drawings that represent objects from typical viewpoints. Drawings of a horse from the side, for instance, are preferred over drawings that present a horse from above. We prefer images of animals how we typically see them in the world – upright and on their feet (Palmer, Schloss and Sammartino 2014, 210–211).

The basic idea behind both the arousal and prototype theories is that aesthetic preferences are the products of general information-processing systems. This idea is more explicit in *processing fluency* theory, which asserts that the important factor in aesthetic experience is how easy something is to perceive or understand. Objects differ in terms of processing fluency, and this fluency is hedonically marked, in that different levels of positive feelings are associated with different degrees of fluency. Stimuli that are easy to process produce positive feelings and are pleasurable. This positive affect then feeds into judgments of beauty.

There are two processes here: *perceptual fluency* – ease of processing input to the perceptual systems – and *conceptual fluency* – ease of processing conceptual information. Symmetry, high figure–ground contrast and rounded shapes are easily processed visually, and are generally favored over asymmetry, low figure–ground contrasts and jagged shapes. Simple, easy-to-understand ideas are more conceptually fluent than complex, unfamiliar ideas. According to this view, high fluency is pleasurable because it indicates that things are familiar, and that perceptual and cognitive processes are going smoothly. Conversely, low fluency would indicate that there is something problematic in the perception or understanding of a stimulus (Reber 2014).

Not all data seem to support this view, though. Nonexperts seem to prefer pictures and music that are simple and easily processed, while experts tend to prefer more complicated and less easily processed artworks (Reber 2014, 236). If so, it cannot be that simple fluency is responsible for preferences. Perhaps the pleasure we get from aesthetic experience is also affected by expectations. Stimuli that we don't expect to have high fluency, but in fact do, might produce especially strong feelings of pleasure. This may be why an educated listener might prefer more complicated music. Bach's Fugues, for example, may be preferred over the simple song *Twinkle, Twinkle Little Star*, even though the latter is simpler and easier to process. The former, but not the latter, surprises us when we come to process it easily, producing an extra jolt of pleasure. Perhaps this is why some people prefer a more challenging painting from Picasso over the more easily processed paintings of naïve realists. We expect a challenge in viewing a Picasso, and so when we come to understand the painting – we process it – we get more satisfaction. Another principle might be at work here as well. There is some indication that fluency does not have the same positive effect when we know the source of fluency. We know the source of fluency in a too-simple song, or one that we have heard too many times, and that negatively affects our response (Reber 2014, 228–232).

One virtue of the processing fluency theory is that it can potentially explain prototype, exposure and arousal effects. We get more pleasure from the most typical examples of paintings, because they are more easily processed. We prefer artwork that we have already been exposed to, because that artwork is more easily processed perceptually. And we prefer artwork that arouses us because of its collative properties, unexpectedness for instance, when the processing is easier than what we expected. Processing fluency theory also takes into account experience and training, both in how they affect processing and in the expectations that they generate. This can potentially explain why people have different experiences and preferences based on what they are familiar with, what training they have and how that affects expectations.

But processing fluency cannot be the whole story. Aesthetic experience seems to be affected by content in ways that cannot be reduced to mere processing fluency and expectations. Compare the viewing of a portrait of a person we admire with the viewing of a portrait of someone we despise. And suppose the processing fluency of both images is equal. If knowledge and memory play a role in the experience of art, as virtually everyone seems to admit, our affective responses to the two equally easily processed paintings would surely be different. Moreover, *disfluency* may play a positive role. Rolf Reber argues that this is the case in paintings like J. M. W. Turner's *Snow Storm*, where the chaotic brushstrokes, which reduce fluency, represent the chaos of the storm and have positive value (Reber 2014, 236–237). Perceptual disfluency might also function to generate perceptual exploration, as a cubist painting might in representing space as fractured and disjointed, by causing us to look more closely to understand how space can be seen in this way. Similarly, conceptual disfluency (or perhaps "semantic" disfluency) might also stimulate an exploration, as we see in the novels, plays and films that play off ambiguity to generate uncertainty about what happens and what it means. George Wilson argues this is often the case in film and found in classic Hollywood movies such as Nicholas Ray's *Rebel Without a Cause*, where there is both an obvious surface commentary on the social problems of the day, and a less obvious "oblique" commentary on the source and nature of social bonds (Wilson 1986, 166). Understanding both points of view is clearly more difficult than the simple understanding of the obvious surface commentary. Here disfluency might be a virtue.

4.4 Laws of Aesthetic Experience

The basic idea underlying these general theories of art experience is that there are regularities in the functioning of perceptual, cognitive and reward systems in aesthetic experience that can be seen as law-like. It may be, though, that no single, general law-like principle will do. Vilayanur Ramachandran and William Hirstein argue for eight laws or principles of artistic experience, related to peak shift effect, isolation, grouping, contrast, problem solving, unique vantage points, metaphors and symmetry (Ramachandran and Hirstein 1999, 33–34). The peak shift effect is based on the exaggeration of "essential" features of an object that serves as super stimulus: "what the artist tries to do (either consciously or unconsciously) is to not only capture the essence of something but also to amplify it in order to more powerfully activate the same neural mechanisms that would be activated by the original object" (Ramachandran and Hirstein 1999, 17). So a caricature of a face, for instance, would more powerfully activate the relevant neural networks than would a mere photograph. And a sculpture or painting of a female nude that exaggerates some distinctively female shape and posture would more powerfully activate the relevant neural networks than it would if it represented more normal female shapes and postures.

A second law is that the isolation of single cues, such as color or line, may generate pleasure (Ramachandran and Hirstein 1999, 33). A painting, for instance, can give us more pleasure by drawing attention just to color by manipulating it without manipulating line or form. Third are the grouping principles whereby color, for instance, might make us perceive unconnected blotches of paint as representing a single object and provide some perceptual pleasure on that basis. Other laws are based on the ideas that we pay special attention to high-contrast areas; we like perceptual problem solving when a stimulus is not easily interpretable; we abhor unusual or unique vantage points; we enjoy metaphors; and we cue in to symmetry because many of the most important things in our environment, predators for instance, exhibit symmetry in some form.

Surely there are some insights here. At least some of these principles are likely at work in our aesthetic experience of art. And at least some of them are also exploited by artists, as Semir Zeki claims when he argues that visual artists are in effect neuroscientists, investigating the principles that govern perception (Zeki 2000). But there are also some concerns. First, there seems to be more than a hint of selection bias in the examples that Ramachandran and Hirstein give. They are happy to give examples of peak shift effect, for instance, in the sculptures of the female body form. But merely giving a few examples does not confirm it as a general law. What about the photos or paintings that seemingly do not exaggerate anything but are still beautiful or emotionally powerful by virtue of content? Not all the artwork we enjoy seems to be caricature. The portraits of Vermeer, for instance, are hardly caricature, but nonetheless give great pleasure to many people. If so, how can it be a general law? We can recognize that *sometimes* peak shift effect is employed for aesthetic purposes, but that does not require that artwork employs it or that it always functions in a particular way. And sometimes Ramachandran and Hirstein simply make assertions without any evidence at all. For instance, in their discussion of contrast, they tell us that "a nude wearing baroque (antique) gold jewelry (and nothing else) is aesthetically more pleasing than a completely nude woman or one wearing both jewelry and clothes," but they offer no evidence for this generalization, or even that anyone other than themselves has this view (Ramachandran and Hirstein 1999, 27).

But most importantly, this approach is limited by the fact that it does not seem to take environment into account. The environment will be important for two reasons. First, it affects the development of sensory systems, which in turn affects perception in ways that may affect aesthetic experience. As we have already seen, there is good evidence that the ability to hear phonemes may depend on what language is learned. Second, the environment provides the context for our engagement with art. We practice our arts in engineered environments or "niches" that guide our behavior and affect our experience. Both of these factors imply that human art-related behaviors and experience will be variable. How much of this variability can these laws accommodate? We look more closely at both of these factors in the next section.

4.5 The Neurobiology of Aesthetic Experience

In the previous section, we looked at general theories of aesthetic experience based on arousal, prototype effects and processing fluency. Perhaps we can instead start to think about aesthetic experience by looking not at patterns of aesthetic preferences, but at the underlying neurological structures. A full account of these structures is not possible here, but we can understand this approach by looking at how features of the visual system might work in the perception of visual art; the possible role of mirror networks in aesthetic experience; and how reward networks might be engaged in the experience of art.

Neuroscientist Margaret Livingstone, in her book *Vision and Art: The Biology of Seeing*, describes numerous ways the structural features of the human visual system might affect our experience of art. One effect is a product of retinal structure. When we focus on something in a painting, we use the highly packed, color-sensitive cones in the fovea. But if we focus elsewhere, and see that thing only in our peripheral vision, we use the less densely packed and color-insensitive rods of the retinal areas outside the fovea. Foveal vision, which functions in identifying objects and people, is more sensitive to colors, edges and fine details. Peripheral vision, in contrast, is more sensitive to luminance and coarse-grained features such as overall patterns of light and dark. This retinal structure implies that as we scan an image and our focus changes, the features of an artwork can seem to change, depending on whether they are processed in the fovea or in the other parts of the retina.

Livingstone illustrates this phenomenon with her analysis of Leonardo's *Mona Lisa* and her famously ambiguous smile, which seems to fade and reappear. When we focus on Mona Lisa's smile, we use our central foveal vision, which is sensitive to the lines and fine detail of the painting. But if we focus elsewhere in the painting, and see Mona Lisa's expression in our peripheral vision, we see the coarser components, in particular the overall shading and contours of her face. What makes her expression ambiguous, according to Livingstone, is that in the lines and finer details the smile is less pronounced than in the coarser shading and contours of her face. So we look directly at her smile and the smile seems less pronounced. Then we look elsewhere and it seems more pronounced. This generates a conflict and a tension in our visual experience (Livingstone 2002, 68–73).

In Section 2, we looked briefly at how the neurological "mirror" systems might function in the experience of dance. As we saw there, mirror systems seem to operate in both the observation and the execution of particular actions. In an experiment described in Section 2, ballet dancers had more activation in their mirror systems when watching the ballet actions they were trained in than when watching the capoeira actions for which they were not trained. And capoeira practitioners had more activation in their mirror systems while watching the capoeira actions they were trained in than while watching the ballet actions for which they were not. This activation of the mirror systems, Barbara Montero argues, might generate a sort of kinesthetic empathy that would affect our aesthetic experience, based on the simulation of our own proprioception of the actions we observe that we have done before. If so, we might experience aesthetic properties of dance actions we have learned, in ways that are not accessible to those who have not trained in this way.

But if mirror systems function in this way in dance, wouldn't they function similarly in other arts? We would expect that the experience of a musician watching another musician play would involve this same system. A violinist knows what it feels like to play the violin, and a pianist knows what it feels like to play the piano. The violinist and the pianist would then have access to proprioceptive memories not available to those who have not learned to play these instruments. Similarly, a trained singer would likely have distinctive proprioceptive memories that result in kinesthetic empathy when seeing and listening to an opera aria. There might even be similar mirror neuron responses that generate kinesthetic empathy in the brushstrokes we see in a painting, or the pencil marks in a sketch. And surely there would be extensive mirroring activation in our experience of the dramatic arts – plays, television shows and movies. The actions we see represented are often actions that we have executed ourselves. And the emotions that we see are typically emotions that we feel ourselves. If mirror systems are activated by the observation of action and emotions, there will surely be some sort of empathetic response (Clay and Iacoboni 2014, 315–316).

Moreover, there is some reason to think that when we *read* about actions in a novel or play, we are also engaging mirror systems. In one experiment, for instance, when subjects were asked to silently read sentences that describe actions of the mouth, hand and foot, their premotor cortex was activated in areas that correspond to the relevant body part. Mirror systems seem to function similarly in the description of particular emotions, and the experience of emotions, generating an emotional empathy (Clay and Iacoboni 2014). If mirror systems operate systematically this way in response to spoken or written words, it is not surprising that we can form strong bonds with fictional characters. In some sense, we may be feeling what we read they are feeling, and feeling as if we are doing what we read they are doing.

Some researchers believe that mirror systems also play an important role in the experience of music through motor representations and connections to the emotion and reward systems in the brain. According to Katie Overy and Istvan Molnar-Szakacs, the experience of music is more than just listening to an auditory stream. It is also associated with motor actions: hand and arm actions in drumming, piano and guitar playing and vocal cord, tongue and mouth actions in singing. And we do all these things in groups, imitating each other and synchronizing our actions. Overy and Molnar-Szakacs call this "shared affective motion experience," arguing that "musical sound is perceived not only in terms of the auditory signal, but also in terms of the intentional, hierarchically organized sequences of expressive acts behind the signal" (Overy and Molnar-Szakacs 2009, 492). If so, we experience music, not just as a sound stream but also through mirroring as a product of the actions, emotions and intentions that produced it. And just as we in some sense feel through our mirror systems the emotions we observe in other people, we also feel, through our mirror systems, the emotions we recognize as being expressed in musical action – the playing of instruments or the singing of songs (Overy and Molnar-Szakacs 2009, 495).

What we know about neurobiology can help us better understand the mechanisms that generate emotion and pleasure. The experience of music has been shown to affect neural activity across the brain, but particularly in the amygdala, which is involved in the initiation, detection, maintenance and termination of emotion, along with the hippocampus and other brain structures involved in the experience of emotion (Koelsch 2010, 2014). Neuroimaging has shown musical activation of reward systems, in particular the nucleus accumbens, which receives input from the amygdala and hippocampus, and is also activated by sex, drugs and chocolate. This structure, which is involved in the experience of pleasure, also projects into motor-related brain structures and might generate a drive to move to music (Koelsch 2010, 134). Perhaps the activation of these reward systems connected to motor areas is the basis of human tendencies for rhythmic entrainment to music, from toe tapping, clapping and head nodding to what seems like full-fledged dance.

4.6 Conclusion

This brief sketch of the psychology and neurobiology of art behavior and aesthetic experience is incomplete and speculative. Not everyone thinks that mirror systems can do all of the things claimed. (See Hickok 2009 and Sapolsky 2017.) But this sketch also shows us how a biological framework might help us understand why and how we engage in the arts. First, it shows us how aesthetic experience of art might be a product of the interaction of multiple systems. The foveal-peripheral visual systems, for instance, might result in conflicting visual experiences that generate interest and perhaps pleasure. Second, it shows us how our aesthetic experience of art might systematically engage mirror systems and produce action and emotional and kinesthetic empathy. Third, it reveals how deeply emotional our aesthetic experience is, and how it produces pleasure by using the same reward systems engaged by other activities.

These insights can be incorporated in the evolutionary thinking of the previous section. It is clear that the mechanisms by which we experience art are not exclusive to our art experience, but are based on preexisting structures that have evolved for other purposes. The retinal structure that affects our experience of paintings, for instance, long predates our visual art. Likewise, the postulated mirror systems that might function in dance, theatre and music predate those activities. And the emotional and reward systems are likely very old in evolutionary terms. In this sense we can think of art as partly a by-product of these systems. But this does not rule out genuinely adaptive value for the arts. The engagement of the mirror systems might, for instance, support the social functioning of music and dance, in enhancing cooperation and promoting social cohesion through action and emotional empathy and coordination.

But there are some additional complications here as well. As we have already noted, along with the uniformity in the psychology and neurobiology among humans, there are also differences. Evolutionary thinking recognizes and relies on genetically based individual differences within populations. Without these differences there could be no adaptive evolutionary change. Second, there will also be variability because of environmental differences. This may occur because development is affected by the environment in which it occurs. We will have different aesthetic experiences depending on what we were exposed to in development. Moreover, the art behaviors and experience depend on the features of the environment in which it occurs. We can only dance tango or ballet, for instance, in an environment in which there is tango and ballet. The next section addresses this environmental component of our engagement with art.

5 The Ecology of Art

5.1 Introduction

Perhaps the most obvious difficulty for a biology of art is that our art behaviors and aesthetic experience are so variable that it isn't clear they can be fully explained in terms of some evolutionary theory, by any laws of aesthetic experience or by common, underlying neurological mechanisms. The evolutionary theory that music and dance function to promote social cohesion and cooperation makes sense, for instance, if we think about church choirs and group folk dancing in the village square. But we also dance and sing solos onstage and in videos. We sometimes attend plays and watch fiction films and thereby learn about human nature and the situations we might find ourselves in, and how to cooperate and avoid conflict. But sometimes we go to outrageous and silly comedies that are so far removed from the world we live in that it is hard to see what insights could be gained. Some people might prefer high processing fluency in their artwork and music, but some people prefer artwork and music that are more challenging. The peak shift effect might sometimes bring pleasure in the artworks that are based on caricature, but sometimes more realism in representation instead gives us satisfaction.

One obvious reason for this variability in how we engage and experience art is that we don't all grow and develop in precisely the same environment. As the interactionist model of development discussed at the beginning of Section 4 suggests, an environment can directly affect development at the most basic epigenetic level by how it affects the expression of genes. There may be relatively little environmentally induced variation at the level of the *primary epigenetic rules* that do things like guide the development of the retina with its distinctive pattern of rods and cones. But at the level of the *secondary epigenetic rules*, which are more sensitive to the environment, we should expect more environmentally induced variation based on exposure to stimuli in the environment.

The influence of the environment extends much further. Most obviously, expertise affects experience, as indicated in the previous section, where experts, who have a greater exposure to art, seem more likely than non-experts to prefer processing disfluency. But also, someone growing up in an environment that gave them exposure to Picasso paintings and Duchamp readymades in fine art museums would likely view painting and sculpture differently than would a 15th-century Italian viewer of religious painting and sculpture, who expected these artworks to function in religious ways. And someone who grew up in the 21st-century dance environment that contains professional ballet, competitive ballroom, Broadway-style tap and jazz and hip-hop for music videos, all of which involve extensive training and seem to be primarily for public performance, would surely view dance differently than would someone who grew up in the 18th century and knew only the ritualistic folk dances performed in their village square by amateurs. Furthermore, the person who grew up listening only to Western popular music might not be engaged by traditional Chinese opera in the same way that a native aficionado would. The question raised here is this: how can a biological account accommodate the full diversity of art behavior and experience? To answer, we look at the *ecology* of art, the study of the interaction of organisms in their environments.

5.2 Engineered Niches

The term "ecology" is relatively new, coined in the late 19th century by Ernst Haeckel, even though ecological thinking has been around much longer, at least since Aristotle, who was keenly interested in how organisms function in particular environments (Stauffer 1957, 138). After Darwin, ecology acquired a new significance in that the interaction of organisms in their environment is a crucial factor in the operation of survival selection. Even so, ecology became a distinct discipline only in the second half of the 20th century (McIntosh 1988, 1). One of the most recent and striking developments in ecology is the turn to systematic thinking about *niche construction*.

Naturalists have long been aware that organisms across biodiversity modify their environments, but only recently have they begun systematically theorizing about it. In a 2003 book, *Niche Construction*, John Odling-Smee, Kevin Laland and Marcus Feldman lay out the basic framework. The foundation of this approach is the recognition that organisms can become adapted to an environment in two ways. First, organisms can be modified evolutionarily through selection processes, as we see in the long fur and thick blubber that has evolved to deal with extreme cold. But organisms can also change their environment to better suit their physiology and lifestyle. It is well known that beavers build dams, elephants clear trees and underbrush, and birds build nests, but there are countless other examples. Badgers dig underground tunnels and nests with nurseries, latrines and sleeping chambers. Earthworms modify the soil they live in to better suit their aquatic physiology. Ants and bees build nests that regulate temperature and humidity. Plants and trees modify the soil surrounding them to prevent competing plants from growing. Bacteria create biofilms. Most strikingly, some termites build large nests that have complex ventilating systems driven by fungus farms, and where they can live in relative safety and comfort (Odling-Smee, Laland and Feldman 2003, 80–84).

We can think of these modified environments as *engineered niches*. If so, humans are impressive niche engineers. Most obviously, we construct the buildings we live and work in – our factories, schools, office buildings, stores and shops. We make the roads and freeways on which we drive our manufactured cars. We eat bread made in factories from engineered grains, and butter and cheese we get from dairy cows bred to more efficiently produce milk. We see this niche engineering in the games and sports that we play in gyms and on courts and fields. And we see it in the arts as well. We dance in clubs, on stages and in ballrooms. We paint in studios and exhibit those paintings in galleries and museums. We sing in churches, at nightclubs, on the opera stage and in huge stadiums at sporting events. And we make and listen to music on electronic devices. In general, our art behaviors take place in highly engineered environments that guide the expression of these behaviors.

We don't interact equally in these niches. Some of us, such as professional ballet dancers, symphony musicians, filmmakers and visual artists, interact extensively in fairly well-defined and narrow art niches, but many of us may interact only sporadically in art niches, perhaps just occasionally singing in a church choir, acting in a school play or visiting an art museum. And those who interact in the dance niches are typically not the same as those who interact in the music niches, who are not the same as those who interact in the visual arts niches. Within each of these broader art niches, there are also more specialized niches. In the dance niche, for instance, are the ballet, modern dance, ballroom, hip-hop, salsa and Argentine tango subniches, populated by people who might occasionally interact with those in other dance niches, but who primarily interact only within a single subniche. In biological terms, there is a hierarchical and demic structure to the ecology of art, where there are groups of individuals interacting at different levels and to various degrees within their respective niches.

5.3 Human Niche Technologies

We can think about our engineering of these niches in terms of technologies. There are the *architectural technologies*, the buildings and spaces in which we interact. In the arts, these are the theatres, museums, studios, concert halls and so on. But there are also the artifacts that we make to help us in arts activities – our dancing shoes, musical instruments, canvases, brushes and paints, video cameras and televisions and the electronic devices that help us record and play music at our leisure. These *artifactual technologies* affect how we engage in art. We can dance "en pointe" in ballets, for instance, only because someone invented pointe shoes. We can paint on canvases and play the violin only because someone made the paint, canvas and violins. We can make films only because someone invented the technologies that made that art form possible.

There are less concrete technologies as well. Within each niche there are the *cognitive technologies*, which include the concepts that guide our behaviors. They tell us what counts as a proper instance of each of these arts, and thereby regulate our activities. To count as doing *dance*, for instance, we must satisfy the conditions associated with that concept (whatever they may be). And to do *ballet*, there are further restrictions. In part, this will involve doing specific actions associated with subordinate concepts – *pirouette, tour jeté, coupé jeté, tendu, développée* and so on. When we learn these subordinate concepts, we learn what is involved in performing each particular action. And while most of these concepts may be found only in ballet niches, there may be some bleeding over into other dance niches. In ballroom dance, for instance, one might also do a *développée* or *chaîné turns*. In the other arts as well, there will be the general concepts of *painting, sculpture, theatre* and *music*. And within each of these more general niches, there will be subniches, each with its associated subordinate concepts. One might play the violin, for instance, in a classical, bluegrass or jazz niche.

The recognition of these concepts leads to *conventions*, practices that guide behaviors. In ballet, for instance, it has become conventional to do *pirouettes, tour jetés* and *coupé jetés*. In other dance forms, other concepts ground conventions. In *ballroom dance* are the subsidiary concepts of *waltz, tango, foxtrot* and *quickstep*, each with its own set of associated and overlapping action concepts and conventions. There are also the epistemic and pedagogical technologies that function in teaching and learning an art form. *Epistemic technologies* tell us how to do a particular action or kind of activity. These include the traditional dance manuals that give explicit instructions for each action, but also the video dictionaries of dance actions that can now be accessed online. *Pedagogical technologies* help us learn how to do the specific actions and more generally the activity associated with the concept. The syllabi in ballet and ballroom dance, for instance, tell us what actions should be learned and mastered at each stage of education, on the assumption that we can better learn the activity if we learn specific actions in a particular order. Other pedagogical technologies include the exercises, classes and procedures for how the classes will be conducted. In ballet, for instance, there is "barre work," which relies on the use of a ballet barre for support, to better facilitate the training of muscles in doing particular actions. This is in contrast to "center work," which does not rely on the barre and is used instead to develop balance. And within each of these types of class work are specific exercises designed to enhance the training. Similar pedagogical technologies have developed in other arts, in the drawing drills and painting exercises of the visual arts, the vocal exercises in singing and the bowing exercises and scales in music.

Alongside these cognitive, epistemic and pedagogical technologies are the institutional technologies – the socially constructed and recognized organizations that function as agents in a variety of ways, and that often have explicit legal status. Universities, art academies, dance conservatories, orchestras and ballet and opera companies all play important roles in how each art activity is conceived, learned and executed. Universities, for instance, confer official degrees that grant social status to practitioners. Concepts, conventions and practices are passed on from those who have mastered them to those who are hoping to learn them. Having a degree or certificate is a sign that one has learned these concepts, conventions and practices. There are also professional organizations that regulate these activities. The Imperial Society of the Teachers of Dancing (ISTD) is one example, and it has divisions for ballet and ballroom dance. It provides training for teachers, and sponsors examinations that, when passed, confer credentials recognized by other practitioners. These institutional technologies might harden the niches by making the boundaries more distinct both in terms of who occupies the niche – those given official recognition, and what cognitive, epistemic and pedagogical technologies are found in each niche – and in terms of which of these technologies are assumed in the testing and credentialing process.

5.4 Niche-Dependent Normativity

In Section 2, we looked at two apparent problems with naturalistic, scientific approaches to the philosophy of art. According to the *normativity problem*, science can tell us how we actually engage with and experience art, but not how we *should* do so. Dance, for instance, may enhance cooperation, but that doesn't automatically imply that we *should* dance to enhance cooperation. And just because we prefer the actual Mondrian grids to experimental variations doesn't imply that we *should* do so. According to the *conceptual problem*, to engage in art requires that we have the appropriate concepts. In order to dance, for instance, we must have the concept of dance and act on the basis of that concept. Science can tell us about how the brain functions in our art behaviors and experience, but it cannot tell us about the concepts we use when we do so.

In Section 2, we didn't have the resources to address these problems, but here, armed with the ecology of art, we do. The obvious response to the conceptual problem is that the relevant concept governing a behavior need not be in the mind, but can instead be a feature of the niche. We act, for instance, in accordance with the concept of *dance* in general, and *ballet* in particular, simply by learning the subsidiary technologies associated with that art form. We can dance ballet because we learn how to do *pirouettes, tour jetés, coupé jetés* and *tendus*, not because we have some general concept of *ballet* or *dance* in our minds.

To see how the ecology of art can respond to the normativity problem, we can start by recognizing that normativity is generated by the cognitive technologies and the concepts that govern our actions. When we learn the concept of *ballet*, for instance, we learn something about how to *do* ballet by learning the subsidiary concepts – *pirouette, tendu* and *jeté* – that regulate particular actions. These we can learn through epistemic and pedagogical technologies – ballet classes with their exercises, as well as videos and books. As we learn these concepts and conventions, we are also in effect "recognizing" them (to borrow a term from John Searle, 2010, 8). This *recognition* is not just the understanding of a concept, convention or practice but also an *acceptance* of that technology as regulating activities. One can accept a technology as regulating behavior, though, without also approving of it. I might accept as legitimate, for instance, a particular way of doing an action in the Cecchetti ballet syllabus, while still believing it is not the best way to do it.

If those who interact within a niche recognize and therefore accept the cognitive, epistemic and pedagogical technologies in the same way, and they also recognize that others in the niche also recognize and accept these technologies, then there is a *collective recognition*. Each individual accepts the prescribed ways of doing things *on the assumption that others do so as well*. This makes a *collective intentionality* possible (Searle 2010, 42–60). The individual dancers in a ballet company can together perform a ballet and the individual musicians in an orchestra can together play a symphony. There is collective intentionality in these cases when we recognize the same concepts, conventions, practices and implicit values and act on the assumption that others do as well.

This collective recognition of the cognitive, epistemic and pedagogical technologies, and the resulting collective intentionality, generates what Searle calls "deontic powers" – rights, duties, obligations, requirements, permissions, authorizations and entitlements (Searle 2010, 9). *Deontic powers* tell us what we *can* do, what we *must* do and what we *must not* do in particular situations. In performing a traditional *Swan Lake*, for instance, there are options that one *may* or *may not* employ (particular steps in a solo), but there are also things that one *must* do (point one's feet), and things that one *must not* do (a tap dance).

Collective recognition of these deontic powers also generates an enforcement mechanism, *background power*, which is the social pressure to conform to the recognized practices and conventions under threat of negative sanctions imposable by any member of the community (Searle 2010, 160). This background power generally serves to preserve the traditional ways of doing things, and can be brought to bear against those who challenge them. The 1913 production of Stravinsky's *Rite of Spring* choreographed by Vaslav Nijinsky is a famous example, where the violations of conventions by Stravinsky's music and Nijinsky's choreography were reported to result in a near-riot by the audience. These violations of convention can then themselves become conventions, as they get collectively recognized, and generate their own deontic powers. *Rite of Spring*, in various choreographic versions (including Joffrey Ballet's reconstruction of the original) is now in the repertoire of many ballet companies.

Collectively recognized concepts, conventions and practices are also sometimes enforced by an *institutional power* – the power institutions have to enforce rules and regulate behavior (Searle 2010, 141–142). This power, generated by institutional technologies such as universities, conservatories and certifying organizations, can confer status that legitimizes one activity but not another, and one way of doing things but not another, perhaps by how they confer status on the individuals who engage in art-related activities. Universities, for instance, can enforce interpretation of the regulating concepts, conventions and practices by conferring degrees only on those individuals who demonstrate knowledge of the relevant cognitive and pedagogical technologies, and the willingness to accept the institutional interpretations. This power can also be wielded by the institutions that sponsor the arts and the restrictions they place on sponsorship. The US National Endowment for the Arts, for instance, may place restrictions on what can be done, how it can be presented and for whom it is presented. A dance grant could be based on some implicit concept of dance, on particular conventions and on assumptions about proper pedagogical technologies.

There are then, three sources of niche-dependent normativity: first, the deontic force generated by collective recognition and intentionality; second, background power; third, institutional power. We can think of this normative framework as providing us *niche-dependent reasons* to act in particular ways within each art niche. We have reasons to act in conformity with the cognitive technologies, pedagogical technologies and conventions because we recognize and accept them. We have reasons to act in conformity with background power insofar as we care about the opinions of others and how they can affect us. And we have reasons to act in conformity with institutional power because of how it can confer or withhold the status we desire, and how it can sanction us when acting otherwise. We can think of the collective set of these reasons operating in any niche as the *normative situation* – the set of niche-based incentives that guides art behavior and experience.

This niche-based framework of normativity shows how the evaluation of specific artworks might work. There are typically multiple and sometimes conflicting norms within each niche relating to a particular artwork, and consequently multiple and sometimes conflicting reasons to do things one way or another. An artwork may therefore be "good" relative to some concept, norm or convention, but "bad" relative to another. In ballet, for instance, there are norms associated with the use of feet, leg turnout and arms, but there are also norms associated with the choreography and staging of the work, norms that might favor innovation. So a production of *Rite of Spring* (such as the 1913 Nijinsky original) may be bad insofar as the dancers violate the accepted norms about the use of feet, turnout and arms, but it may also be good *on the basis of* these violations, in the sense that the violations of the foot and arm norms constitute a choreographic innovation and make the production better on those grounds. In this case the satisfaction of one norm makes the satisfaction of another more difficult. There may also be other norms related to the body types of the dancers, the staging and scenery, the way the music is presented and so on.

The evaluation of an artwork relative to an engineered niche will therefore typically be heterogeneous, based on different components of the niche-dependent normativity. This makes an overall evaluation of a work of art difficult. It isn't obvious that there is a single metric that is comparable across all norms. How, for instance, can we measure the relative failure to satisfy the norms associated with the use of the feet, turnout and arms in *Rite of Spring*, and weigh it against the satisfaction of the choreographic innovation norm? This problem may be partly solved by evaluative conventions. In some contexts, such as choreography festivals, informal and unstated conventions might weigh choreographic norms more heavily than dance technique norms. There might also be explicit evaluative technologies in a niche that tell us how to weigh multiple norms. In its sanctioned ballroom competitions, the World DanceSport Federation (WDSF) currently mandates the equal weighting of four norms: "technical qualities," "movement to music," "partnering skill" and "choreography and presentation." Each couple is given a numerical score according to each norm, and the scores are then added to provide a total score, which determines the outcome of the competition.

5.5 Niche-Independent Normativity

Alongside the niche-dependent normativity, there is also a *niche-independent* normativity that generates reasons to act, conceive, interpret, experience and evaluate art in ways that are not directly a product of an art niche. This niche-independent normativity is generated in the actual creation and experience of art, based on our preferences and goals, and the pleasure or enjoyment we get from the art. Unlike the socially grounded and enforced niche-dependent normativity, this normativity is individual and personal. It is implicit in the preference studies briefly discussed in Section 2, and the attempts to find the laws of aesthetic experience in Section 4. The basic idea is that we have reasons to engage in particular art behaviors and have particular aesthetic experiences because of the rewards we get from these behaviors and experiences. If we enjoy the actual Mondrian geometric paintings over the experimental variations, as described in Section 2, then we have a reason to experience the prior over the latter. If we find interest or pleasure in the arousal generated by the mismatch between incoming information and our expectations, as Daniel Berlyne speculates, then we have a reason to experience art that generates this arousal. If prototypes give us more pleasure, perhaps by being easier to process, we have a reason to seek out prototypical art. And if dancing gives us a distinct kind of social pleasure and enhances our cooperation, we thereby have reasons to dance.

The discovery of the neurological mechanisms discussed in Section 4, which operate in art behaviors and experiences, can help us understand this normativity. We can understand why, for instance, we have interest in and get pleasure from some aesthetic experience by discovering how the foveal and peripheral systems interact in the perception of Leonardo's *Mona Lisa*. In general, we can understand how aesthetic pleasure is generated by how the sensory and cognitive systems connect with the reward systems. Similarly, we might understand why we can have a particularly rich experience in watching dance steps we have learned, through the operation of mirror neuron systems and the kinesthetic empathy they generate. The point is this: as we learn more about the individual and personal satisfaction we get in our engagement with art, whether through psychology or neurobiology, we learn more about the reasons we have to engage with it in particular ways. What we learn may then generate new and more satisfying preferences. If we understand the reward mechanisms, we may then seek out art behaviors and aesthetic experiences that *better* engage these mechanisms and are *more* satisfying than those we currently have.

This niche-independent normativity is not entirely independent of niches, however. Insofar as we accept the normative situation of an art niche, we might also come to prefer the artworks favored by the norms of that niche, and get pleasure from them. Moreover, the environment can also influence what causes arousal in us, by affecting our expectations, and what kinds of processing fluency we have based on previous exposure. Finally, the associations we form with a work of art surely depend on the circumstances in which we encountered that work. Nonetheless, the pleasure and satisfaction we get from art are personal and not *just* products of the conventions and norms of the niches in which we experience art.

5.6 The Two Streams of Normativity

By recognizing these two streams of normativity, niche-dependent and niche-independent, we can get a better understanding of our evaluation of art. Not only can there be a conflict in the reasons generated by different niche-dependent norms, there can also be a conflict between the niche-dependent and niche-independent reasons. Niche-dependent normativity may incline us in one way, given that an artwork satisfies niche-dependent norms, while the niche-independent normativity may incline us in another way, if that artwork gives us satisfaction or pleasure that is independent of and contrary to the niche-based norms. So we might judge Duchamp's *Fountain* good with respect to niche-based norms related to innovation and conceptual content, but bad relative to the pleasure it fails to give us.

This conflict in normativity suggests that there are grounds by which we might criticize a particular niche based on a conflict of interest. An art niche will typically favor the preferences, capacities and tendencies of some people, but not those of others. For instance, a visual art niche that favors novelty over representational skill and technique will benefit those who are best at generating novelty, along with those who have interest in and get pleasure from novelty. But it will not benefit equally those who have capacities for and interests in and who get pleasure from representational skill and technique. A particular art niche might also have norms that favor less-educated over more-educated listeners. Or a niche might favor those with a particular political view over those with other views. Or a niche might favor one kind of person in terms of nationality, culture, race, gender, body type etc. over others. Or a niche might favor artists over audiences – if artists have niche-dependent reasons to produce artwork not favored by typical preferences and capacities of audiences. Finally, a niche might even favor critics over artists or audiences – if the critical discourse about art is valued over the experience or making of art.

Within a niche there may also be a conflict in interpretation. While we can only gesture toward the factors here, we can see that within each niche there might be multiple interpretative technologies. There might be conventions for interpreting according to the creator's intentions, or according to reader/viewer response. There might be norms for interpreting according to semantic conventions analogous to language. We might interpret on political grounds or in terms of gender and race relations, economic classes or religious commitment. If so, the interpretation of a particular artwork might "mean" something relative to one convention, and it might "mean" something else according to another. And there may be a further conflict between these niche-dependent interpretive technologies and niche-independent tendencies. There might be a natural tendency to treat artworks as intentionally produced artifacts to be understood in terms of the intentions of the creators. If so, then a particular niche that gives reasons to ignore these intentions will conflict with the natural tendencies of the audience.

We can now better understand the source of the normativity problem laid out in Section 2, that science can only give us the facts about our art behaviors and experience, but not the "shoulds." George Dickie, who made this argument, also has claimed that we should understand our engagement with art through the "rules of the art game," the practices and conventions associated with art, whereby normativity is generated by these rules (Dickie 1962, 299). But now we can see that *of course* a niche-independent personal preference doesn't by itself and automatically imply a niche-dependent "rules of the game" kind of "should." These are two sources of normativity – two ways art can be good or bad. There is, however, more to understanding art-related normativity in understanding the nature and source of the value of art – in its "mattering." We begin the next and final section there.

6 Conclusion

6.1 The Value of Art

This is an Element about the biology of art, but many of the topics addressed here are philosophical in nature and require philosophical treatment, in particular two of the most fundamental questions: What is art? What is its value? In Section 1, we looked at attempts to answer the first question, but concluded there that none of them seemed fully satisfactory. We return to that question later in this concluding section. We have made some progress in answering the second question about the value of art. The proposed functional definitions of art in Section 1 focus on the valuable things that art can do. It can produce pleasure and aesthetic experience. It can express an idea or an emotion. It can engage the imagination. In Section 3, we saw that art might serve some evolutionary function, perhaps in enhancing cooperation or in providing insights into the situations we might find ourselves in, and a set of strategies for dealing with them. Or it might serve as a kind of fitness display that is advantageous in reproduction. In Section 4, we looked at how art might get its value through arousal, by processing fluency and by engaging our reward systems, based on neurological connections with our sensory, motor and memory systems. And in Section 5, we saw how engineered niches guide our art behaviors and generate normativity through cognitive technologies, conventions, norms and background and institutional power.

All this seems to show that there are many ways art *might* have value. But it does *not* show that all art has some particular value in all cases – that there is some single value to all art. One problem with the functional definitions of art, you might recall, is that some art sometimes seems to have the proposed functional trait, perhaps in generating pleasure or aesthetic experience, or in expressing an emotion, but not all art does. Some widely recognized works of art are disturbing and unpleasant, some don't seem to generate aesthetic experience, some don't seem to express anything and so on. The question, then, is this: given these diverse ways art can have value, how should we understand the value of art in general?

One philosophical starting point for thinking about value is in terms of language – how we talk about artistic value. When we say something is good, as in "the painting is good," we seem to be attributing the property of *being good* to the painting. Usually it is assumed that this goodness of an artwork is in turn based on the other properties it has. This implies that we can understand the goodness and value of art by understanding the properties of art – what properties good artworks have and what regularities there are in this goodness. Then we would presumably know what gives value to art (Richards 2004). This idea is lurking behind the psychological studies of art in Section 4 that looked for the grounds of aesthetic experience in the properties of art that cause arousal or that make it easily processed, which then generates interest or pleasure. This way of thinking about value in terms of properties can perhaps also account for the functional definitions of art, discussed in Section 1, that try to define art in terms of the valuable things art does. Value would then be generated by the properties of artwork that are responsible for the expression of emotion, the production of aesthetic experience or the engagement of imagination. A song might be good, for instance, because its sound contours are expressive; a painting might be good because its stylistic unity enhances processing fluency that generates reward.

One advantage to thinking about goodness as a property of an artwork is that we can potentially compare the goodness, and hence the value, of two artworks by simply examining and comparing their properties. If one song has the properties that make it expressive to a greater degree than another, then it has more value. But there is also a real problem here, in that the supposed good-making properties of an artwork don't always make the artwork better (Richards 2004, 263–264). Stylistic unity, for instance, can make an artwork worse if the style itself is bad – it is unsophisticated and garish. Or being expressive might make a song worse if what is expressed is trite and banal. This was also one of the problems with the psychological approach described in Section 4 that looks for value in the properties that enhance processing fluency. Some people prefer artworks that are more easily processed, but some, particularly those who are more educated, do not. This suggests that it is not just the properties of an artwork that are relevant; the properties of the viewer or listener are also relevant to the value of an artwork.

We might solve this problem, as some have, by conceiving of goodness or value as a *secondary property*, which, unlike primary properties, depends on a subject (Richards 2004, 2017). The shape or size of an object, for instance, is a primary property because it is independent of any observer. The color of an object, on the other hand, depends on the perceptual system of an observer and is therefore a secondary quality. Nothing is red for someone who is totally colorblind. This idea captures the insight that whether an artwork is good or not depends partly on those who experience it. But this attempt to think of value in terms of secondary properties is misleading in that if a secondary property is dependent on an external subject, we are no longer speaking just of the object, but also of a *relation* between an object and a subject. On this way of thinking, something does not simply have value and is therefore simply good. It has value *only* for someone, and is therefore good *only* for someone.

6.2 Relational Value

This way of thinking is reinforced by a naturalistic approach to value, which starts by looking at how value actually gets generated. In general, we can see that art has value because it matters positively or negatively to someone in some particular way and in some particular context (Richards 2004). A religious painting may matter positively to 15th-century Italian believers by generating a particular religious emotion in a particular context. That same painting may matter positively to the viewers in a 21st-century art museum through the pleasure it produces as they contemplate the skill of the artist. Or a painting might matter negatively to someone who finds it ugly, or who finds its content offensive or disturbing, or to someone who associates it with some unpleasant memories. A group folk dance may matter positively to the members of a village who appreciate and value the social interaction it provides. It may matter simply due to the pleasure that entrainment to music generates. But it may also matter positively to tourists who watch it and appreciate the skill of the dancers. The point is this: an artwork may matter in different ways to an individual person in a particular context; it may matter in different ways in different contexts; and it may matter in different ways to different people. Any theory about the value of art must take this diversity of mattering into account.

So instead of thinking about the value of art in terms of the simple predication of a value property, *x is good (or bad)*, we can think of value here as a relation: *w is good (or bad) with respect to x for y in z*. Here *w* is an artwork, a trait of an artwork or an art-related action; *x* is a way of mattering; *y* is a subject or group of subjects; and *z* is a context or set of contexts. This way of thinking about value has two obvious advantages. First, it better represents how value is generated, by mattering to subjects. Second, this schema captures all the factors that are important in the valuation of artworks: the respect in which they might matter, the subject or subjects to whom they might matter and the contexts in which they might matter. (For more detailed analysis, see Richards 2004 and 2005b. For a general account of relational value, see Richards 2005a and 2017b.)

This relational value schema reveals the difficulties in comparing the value of artworks. While we might compare the amount of pleasure that two songs give a particular person or give two different people, it is less plausible to compare the amount of pleasure they give to people and the other ways they can be good. One work of art might produce pleasure for some person, but not be good relative to contemporary evaluative conventions. Another might be good relative to contemporary conventions, but not produce pleasure in a particular group of people. We see this conflict all the time between popular and elite art. We might, for instance, get great pleasure out of the *Saturday Evening Post* illustrations of Norman Rockwell, even though that kind of art did not satisfy the mid-20th-century art conventions that favored abstract art over sentimental narrative art. Conversely, a Jackson Pollock painting might be good by satisfying those conventions without being good in another way, by providing pleasure to some average viewer.

According to this way of thinking, an artwork can have value and be good by satisfying conventions *only if* the satisfaction of conventions matters to the subjects of interest. If the conventions don't matter in any way to some particular person, an artwork cannot have value *on that basis for that person*. This is clear in cases where the conventions are not known and therefore not accepted. I might attend, see and hear a Chinese opera, for instance, but since I do not know the conventions and norms in that engineered niche, and the satisfaction of those norms do not matter to me, it cannot have value for me on that basis. It can, however, still be good for me by providing a satisfying aesthetic experience and pleasure. For an aficionado of Chinese opera who knows and accepts the conventions, on the other hand, this opera may be good (or bad) on niche-dependent grounds – and have value on those grounds.

Notice also, though, that we can deliberate in the abstract how an artwork might satisfy any niche-dependent norms, whether or not anyone *actually* cares about those norms. An art historian might do this by considering how well a medieval painting satisfies the representational norms related to the spatial perspective of the time – even if no one now accepts those norms. Similarly, we could consider in the abstract how that painting satisfies norms related to theological dogma – even if no one now accepts those norms. This is a sort of conditional analysis: on the assumption of some niche-dependent norm, how good is artwork x? In cases like this, we might have several attitudes toward the relevant norm. We might accept it, reject it or simply have no view about its acceptability. The satisfaction of that norm may therefore have positive value or disvalue or not matter at all.

6.3 Sources of Value

This biological account of art suggests there are three main sources of value to art. First is the evolutionary value. Insofar as art enhances cooperation, facilitates the understanding of social situations and human interaction or functions as fitness display, it has functional value – perhaps with survival and reproductive implications. Second is the personal value in the interest and pleasure art generates through aesthetic experience, expression or the engagement of imagination. Insofar as we are engaged, aroused and get pleasure from art it has value for us. Third is the social value generated through the normativity of engineered niches. Insofar as we accept the cognitive technologies, conventions and norms of a particular niche, and they matter to us, art has value through conformity with these norms.

One obvious advantage of this way of thinking about value is that it easily explains the evaluative judgments we see about art. Some art has broad value in that it produces a particular valuable response in subjects. Mozart's music, for instance, satisfies accepted classical music norms, and seems to provide pleasure to many people in a wide range of contexts. But there is also pervasive disagreement about the goodness of particular artworks. We should expect this, given the heterogeneity of value, particularly when the conventions within a niche conflict with the personal value for those who don't accept those conventions. Art that is meant to shock on the basis of conventions within a niche, for instance, may conform to these conventions and thereby have value for those who accept these conventions. But this art may also be disvalued by those who don't accept these conventions, especially if it does not produce the aesthetic experience and pleasure that they do value. Political art may be valued by those who agree with the message and disvalued by those who don't. And at the most basic level, insofar as we vary in our sensory and psychological makeup, we might also vary in what we find interesting and pleasurable. The colors of a painting may produce a valued sensory experience for one person but not another, based on the idiosyncrasies of each person's sensory and reward systems.

This framework suggests there is no single, simple overall value judgment we can make of an artwork or of art in general. First, since value is relativized to subjects and contexts, evaluative judgments cannot be made on the mere possession of some property of an artwork, but must take into account how valuation can vary for different subjects in different contexts. Second, because there is no obvious metric by which to compare the evolutionary value of an artwork, the pleasure an artwork can provide and the value of satisfying an accepted niche-generated norm, these ways of being good cannot be easily aggregated and compared. If so, the attempt to rank artworks according to overall goodness or value is misguided.

The American Film Institute ranked the "100 greatest American films of all time," first in 1998 then in 2007, based on a survey of film artists, historians and critics. As it has in previous years, Orson Welles's *Citizen Kane* tops the list. But it is implausible that such a list could possibly capture, aggregate and compare all the ways these films can be good or bad for all viewers. At best, this might be merely a ranking of films on how well they are judged to satisfy the accepted norms of an engineered niche at a particular time (1998 or 2007), perhaps with some additional influence from the pleasure each film generates in those surveyed. But since niches change, their associated normative situations change as well; the resulting evaluative judgments would likely be subject to change over time. And as we saw in Section 5, we can evaluate the niche-dependent norms themselves, in terms of how they relate to the niche-independent personal value by generating pleasure and interest, or by how they benefit some people over others. The bottom line is that there is no single, unchanging overall standard by which to evaluate art! Value is too messy, complicated and subject dependent.

There is a subjective element here, in that value depends on mattering to subjects, but this is not to say that art criticism is just a matter of opinion. There are objective facts about the technical skill of the artist and the experiences of the viewer and listener. The abilities of Baryshnikov to dance in ways that engage us far outstrip those of most other professional ballet dancers, let alone amateurs or nondancers. We see it in his musicality, leaping and turning ability and stage presence. The abilities of Vermeer to paint people and places that engage us far outstrip those of most other painters, let alone amateurs and non-painters. We see it in his ability to use perspective, his control of the brush and his use of luminance. The abilities of Mozart to produce music that engages us far outstrip those of most other composers, let alone noncomposers. We see it in his mastery of melody, phrasing and harmonic progression.

The abilities of Maria Callas to sing and act in operas outstrip those of most other opera singers. We see it in her musicality, performing ability and distinctive voice. In each of these individuals there are objective facts first about skill and technique that do not depend on anyone's opinion, and there are objective facts about the responses of those who experience their artistry that are also not matters of opinion. Much more needs to be said here, but one distinctive attribute of art may be our instinctive tendency to evaluate it, particularly in terms of the objective, technical skill demonstrated. This is a fact of some significance and that may be relevant to understanding the nature of art, to which we now return.

6.4 What Is Art? Redux

At the end of Section 1, there did not seem to be a satisfactory definition of art based on its obvious properties. Since not all art is expressive in a particular way, and not all art seems to produce pleasure or aesthetic experience, these properties cannot be used to define art and distinguish it from other kinds of things. And the cluster approach that requires just some subset of these properties does not give us guidance about which features are relevant and how many of them are necessary. I suggested there that a more theoretical approach would be helpful. Just as we can better understand what water is through its theoretical definition – as H_2O – than we can by looking at its manifest properties – its wetness, how it tastes and so on – we can better understand art through a theoretical definition based on a biological approach. Since we now have a more complete theoretical framework, can we give an analogous theoretical answer to the question: what is art?

The answer will not be some simple formula like "water is H_2O." Human art behaviors are far too complicated to give such a simple definition. Perhaps we can return to this question with the three biological components examined in this Element: the evolutionary, the psychological and neurobiological and the ecological. The first component, the evolutionary, places art in the bigger picture, as an evolved behavior that distinguishes humans from other creatures but perhaps also shows some continuity. The second component, the psychological and neurobiological, focuses on the personal experience and underlying neurological mechanisms of doing and experiencing art. The third component, the ecological, is social and focuses on how our engagement with art is channeled by engineered art environments with their technologies, conventions and norms.

The problem with evolutionary theories of art, as you may recall from Section 3, is that not all art seems to function relative to survival and reproduction in the same way. Some art may teach us about human relations, social situations and strategies for dealing with them, but some does not. Some art enhances cooperation, but some art does not. Some art seems to function as fitness display and to function in sexual selection, but some does not. In Section 3 we looked at the possibility that particular art forms – storytelling, music and dance – might serve multiple evolutionary functions relative to survival and reproduction, and at the same time also be by-products of other adaptations, based on structures or systems that evolved for other purposes. If so, then while art cannot be defined by *particular* evolutionary functions, we might nonetheless still think about art in evolutionary terms.

6.5 Biotic Art

In Section 3, we also looked very briefly at the claim by biologist Richard Prum that there is a distinct mode of aesthetic evolution, in which art evolved not just in humans, but also in animal courtship displays and in the bright coloration of flowers and fruits. This "biotic art," according to Prum, is a "form of communication that coevolves with its evaluation" (Prum 2013, 813). Bird song, for instance, has evolved the way it has because of how it is evaluated by birds. In some bird species, for instance, females seem to prefer large song repertoires, and respond more favorably to males to the degree they have large repertoires. If having a large repertoire is an advantage in mating, then this evaluation – preference by other birds – will favor larger repertoires (Pfaff et al. 2007). This kind of evolved communication contrasts with alarm calls, which must simply be recognized to function (Prum 2013, 816). This coevolution of communication with its evaluation also occurs in the flowers that attract pollinators. It is not just that a flower has a particular color, which might merely communicate the location of the pollen, but the pollinators also have sensory preferences that lead them to prefer one flower over another. The plant and the preference of pollinators coevolved to produce the beautiful flowers that we see. The beauty of flowers, Prum notes, contrasts with the plain and drab appearances of root systems, which merely function to absorb water and nutrients. *Aesthetic evolution* is the reason flowers are beautiful and roots are not (Prum 2013, 815).

This way of conceiving art is most obviously compatible with the sexual selection theories of art as display, where what is communicated is fitness, which affects reproduction and the evolution of the trait used to communicate. But perhaps it can be applied to other postulated evolutionary functions as well. A story that educates about social roles and responses to social situations, for instance, might be better understood and better facilitate the right kind of learning if it has particular qualities that are preferred by potential listeners. Descriptive language, rhythm and rhyme might be both preferred by listeners and make for better learning. Singing and dancing might make social coopera- tion stronger by communicating willingness to cooperate, and the better the singing and dancing, the better the resulting cooperation. If so, then singing and dancing will coevolve along with their evaluation. In general, the quality of the communication matters here and has evolutionary consequences that affect the evolution of that communication.

Perhaps Prum gets something right in that human art also seems to be subjected to this critical attention, and that could potentially be the basis for its coevolution. When confronted with a song, dance, story or painting, we tend to take a critical stance. How good is it? We also tend to take a critical stance about the creator. How good is he or she? But Prum's proposal extends the idea of art far beyond the standard usage of the term "art." A construal of this term that includes colorful flowers and bird plumage is not how we started this Element. We began with the human arts – music, dance, painting and so on. Can this idea of *biotic art* be so broad but still accommodate what is distinctive in human art? Perhaps we can see how by returning to evolutionary thinking, not in terms of evolutionary processes, but in terms of history.

When we look at the evolutionary tree, we see that this coevolution of communication and evaluation has appeared multiple times, and in many branches of the evolutionary tree. It evolved in the plant branches of the tree that have beautiful flowers and brightly colored fruit. It evolved in all of those branches that produce mating displays, including the birds with brightly colored feathers, birds that decorate their bowers or birds that seem to dance. It evolved in the branches of spiders that seem to dance, gibbons that seem to sing and more. Because these behaviors evolved in different lineages – on different branches of the tree of life – they are different in their manifestation. Each lineage has its own distinctive suite of traits that forms the foundation for any evolutionary change. The development of a system of communication coevol- ving with evaluation *must be* different in plant lineages, insect lineages, bird lineages and human lineages, because each of these lineages is so different.

We see a similar development of the traits *wings* and *intelligence*, each of which has evolved in multiple lineages and in very different ways. We cannot define *wings* in terms of feathers, for instance, because feathers are found only in the bird lineages. Wings developed in very different ways in the bat and insect lineages, based on mammal and insect body plans, respectively. Similarly, intelligence seems to have evolved very differently in the various lineages. Human intelligence is different from that of other primates, which is in turn different from bird, octopus and insect intelligence, each with its own distinctive capacities and tendencies. Perhaps we can think about *biotic art* in this way, as a single trait developing in different ways in each lineage.

What is distinctive about the development of this *biotic art* in the human lineage is first the complexity of the psychology and neurological mechanisms that function in the creation and experience of art. Perhaps other species have aesthetic experience in their engagement with these communication systems, but we should not expect their experience to be identical to that of humans. Humans have more complicated sensory and cognitive systems, so we should expect human experience to be more complex. Second, the degree and nature of niche engineering are also unique in humans, and so we should not expect other species to create highly complex art niches with architectural, artifactual, cognitive and institutional technologies. We shouldn't expect that nonhuman communication systems would be so technology dependent, and with the same conceptual engagement and institutional guidance. So the differences between human singing and dancing and that of birds or spiders would be expected, given the differences in inherited physiology, psychology and environmental engineering.

Perhaps we should also think of art as a capacity and behavior that has evolved to different *degrees* in the lineages in which it has appeared. If so, it has evolved to a much greater degree in the human lineage, to a lesser degree in bird lineages and to a minimal degree in other lineages. And in some lineages it has not developed at all. This is analogous to *intelligence*, which developed to a higher degree in humans, to a somewhat lesser degree in other primates and in whales and dolphins, to a yet lesser degree in birds and other vertebrates and even less in insects and invertebrates. The point is this: it is possible to think of art as a single thing with great variation in how it evolved, and the degree to which it evolved, because we think that way about many things that have appeared and evolved on the evolutionary tree.

But even so, this idea of biotic art is still too broad. There is also *human* communicative display with evaluation in physical beauty and athletic activities. Both seem to communicate something about the physical state of individuals, and likely evolved through evaluation (assuming beautiful and athletic people generally have better chances to reproduce because of their beauty and abilities). Should we then treat physical beauty and athletic ability as art? This may seem implausible, but physical beauty and athleticism certainly play a role in some of our arts. We seem more interested in watching beautiful and athletic people dance or act in the storytelling arts, television and film in particular. But in human art niches, with their cognitive technologies, we don't typically identify *mere* physical beauty and athletic ability as art, even though both might play a role in the arts. Shouldn't we pay attention to what *our* engineered niches tell us about what counts as art? The concept of *art* is, after all, a cognitive technology in human niches. And if we think about *art* this way, we might naturally ask a pragmatic question: what is the value of conceiving art in one way or the other? We cannot, after all, just "discover" what art is by just looking at various things in the world. There is no manifest and unambiguous property of "artness" in the paintings, songs, stories and dances that we can observe. Rather, we must make a *decision* about what counts as art.

6.6 Pragmatism in Conceiving of Art

Prum's biological proposal would require a radical revision in the conception of art, broadening it to include all sorts of phenomena not normally thought of as art. In contrast, conventional approaches to understanding art are narrow, not just in that they take art to be a distinctly human trait but also in that they also tend to begin from a particular disciplinary perspective. For philosophers, for instance, art is usually understood narrowly as something humans do intentionally with particular concepts in mind, and in the context of conventions and art institutions. For art historians, art is typically understood narrowly in terms of changing human traditions. Cultural anthropologists have similarly adopted a narrow conception of art, associating it with human cultural practices and their meanings. Psychologists and neurobiologists also seem to conceive art narrowly and in terms of the psychological and neurological mechanisms that operate in the behavior and experience of art.

For the cynic, there are obvious advantages to these narrow conceptions of art. Philosophers, art historians, anthropologists and psychologists each have a limited set of tools in which they can approach their subjects, and a narrower conception of art may better suit the resources available within a single discipline. Philosophers can adopt a conception of art that allows them to understand it solely in philosophical terms, cultural anthropologists can adopt one that allows them to understand art solely in cultural terms, psychologists can adopt one that allows them to understand art solely in psychological or neurological terms and so on. On pragmatic grounds, then, philosophers, art historians, anthropologists and psychologists might each have good reasons to adopt a narrow conception of art that makes their respective disciplines adequate.

But there are also good reasons to adopt the comprehensive, biologically based conception of art developed here. First is its explanatory power. It provides theoretical and empirical resources that we can use in explaining art-related phenomena, in particular, the similarities and differences in art phenomena. The rhythmic movements of diverse human cultures look similar, and these behaviors also look to a lesser degree to be similar to what we see in the rhythmic movements of sage grouse and peacock spiders. But if dance were nothing more than a cultural practice – as a narrow anthropological approach might suggest – or conceptually based action – as a narrow philosophical approach might suggest – then these apparent similarities are inexplicable. Why would human cultures that are very different exhibit such similar dancing behaviors? And why would we see similar behaviors in other species? But if these behaviors evolved by selection processes to do similar things, or are based on shared neurological mechanisms, then similarities would be expected. This framework also gives us the resources to explain the differences in art behaviors, based on different evolutionary histories and processes, as well as the different environments of development and expression. This framework tells us that we should expect art behaviors to be similar in some ways, but different in others, and it tells us why.

There are other reasons to adopt this approach that I can only gesture toward here. A second advantage of the comprehensive biological framework outlined here is its resources to correct errors. The evolutionary account in Section 3 suggests, for instance, that any philosophical theory that postulates a single function for art is mistaken. Art can function in multiple ways, relative to education, cooperation or fitness display, and it may do so through representation, expression, or the pleasure it brings in motivating these advantageous behaviors. A third advantage of this approach is its fertility – its resources to suggest new questions, areas of investigations and theories. It reveals, for instance, how foveal and peripheral vision might operate in the perception of art and generate a distinctive experience. It suggests that we should look at specific ways mirror systems might function in the experience of art. This approach also reveals the two streams of normativity outlined in Section 5, the niche-independent and niche-dependent, that generate reasons to engage and experience art in particular ways based on psychology and conformity to accepted niche-dependent norms. If so, then we would expect that sometimes these streams conflict and generate conflicting evaluative judgments.

Finally, this framework gives depth to our explanations in that we can better understand art through an integrated framework. To understand some art phenomenon, we can start with the details of the environment and engineered niche, but then turn to psychological preferences and neurobiological mechanisms, all in the light of evolutionary thinking. If we can fill out this explanatory framework, then surely we have a deeper understanding of art. This integration is in the spirit of the approach biologist E. O. Wilson advocated in his 1998 book *Consilience*. According to Wilson, consilience is the connection of causal explanations from across the branches of learning into a single, unified explanation. And while the humanities and social sciences will have a role in this explanation, it is grounded on the natural sciences, biology in particular. Wilson tells us: "The main thrust of the consilience world view ... is that culture and hence the unique qualities of the human species will make complete sense only when linked in causal explanation to the natural sciences. Biology in particular is the most proximate and hence relevant of the scientific disciplines" (Wilson 1998, 267).

Much more empirical and philosophical work is necessary to fill out this biological framework for understanding art. The evolutionary theories remain speculative. Human psychological, sensory and neurological functioning in the experience of art is still only partially understood. And there are many complexities in the ecology of art, about the nature of engineered niches and how they function. Nonetheless, a biological framework can help us see how to integrate all these approaches and how to have a deeper and more nuanced understanding of the thing we call "art."

References

Aristotle (1987) *The Poetics of Aristotle*, trans. S. Halliwell. Chapel Hill, NC: University of North Carolina Press.

Bachner-Melman, R., and C. Dina, A. H. Zohar, N. Constantini, E. Lerer, S. Hoch, S. Sella, L. Nemanov, I. Gritsenko, P. Lichtenberg, R. Granot and R. P. Ebstein (2005) "AVPR1a and SLC6A4 Gene Polymorphisms Are Associated with Creative Dance Performance," *PLoS Genetics* 1.3:394–403.

Bacon, F. (1960) *The New Organon and Related Writings*, ed. F. H. Anderson. Indianapolis, IN: The Bobbs-Merrill Company, Inc.

Beardsley, M. C. (1982) "Redefining Art," in *The Aesthetic Point of View: Selected Essays*, pp. 298–315, ed. M. J. Wreen and D. M. Callen. Ithaca, NY: Cornell University Press.

Bell, C. (1914) *Art*. London, UK: Chatto & Windus.

Berlyne, D. E. (1971) *Aesthetics and Psychobiology*. New York, NY: Appleton-Century-Crofts.

Boyd, B. (2005) "Literature and Evolution: A Bio-Cultural Approach," *Philosophy and Literature* 29:1–23.

Calvo-Merino, B., D. E. Glaser, J. Grezes, R. E. Passingham and P. Haggard (2005) "Action Observation and Acquired Motor Skills: An fMRI Study with Expert Dancers," *Cerebral Cortex* 15:1243–1249.

Calvo-Merino, B., D. E. Glaser, J. Grezes, R. E. Passingham and P. Haggard (2006) "Seeing or Doing? Influence of Visual and Motor Familiarity in Action Observation," *Current Biology* 16:1905–1910.

Carroll, J. (2011) *Reading Human Nature: Literary Darwinism in Theory and Practice*. Albany, NY: State University of New York Press.

Carroll, N. (1988) "Art, Practice and Narrative," *Monist* 71:140–156.

Carroll, N. (1990) "Interpretation, History and Narrative," *Monist* 73:134–166.

Carroll, N. (2014) "The Arts, Emotion, and Evolution," in *Aesthetics and the Sciences of the Mind*, pp. 159–180, ed. G. Currie, M. Kieran, A. Meskin and J. Robson. Oxford, UK: Oxford University Press.

Carroll, N., M. Moore and W. P. Seely (2014) "The Philosophy of Art and Aesthetics, Psychology and Neuroscience: Studies in Literature, Music, and Visual Arts," in *Aesthetic Science: Connecting Minds, Brains and Experience*, pp. 31–62, ed. A. P. Shimamura and S. E. Palmer. Oxford, UK: Oxford University Press.

Carroll, N. and W. P. Seeley (2013) "Kinesthetic Understanding and Appreciation of Dance," *Journal of Aesthetics and Art Criticism* 71.2:177–186.

Clark, K. J. (2016) "Naturalism and Its Discontents," in *The Blackwell Companion to Naturalism*, pp. 1–15, ed. K. J. Clark. Chichester, UK: John Wiley & Sons Inc.

Clay, Z. and M. Iacoboni (2014) "Mirroring Fictional Others," in *The Aesthetic Mind: Philosophy and Psychology*, pp. 313–329, ed. E. Schellekens and P. Goldie. Oxford, UK: Oxford University Press.

Collingwood, R. G. (1945) *The Principles of Art*. Oxford, UK: Oxford-Clarendon Press.

Croce, B. (1938) "Aesthetics," in Encyclopedia Britannica, 14th edition pp. 263–269, Chicago, IL: Encyclopedia Britannica.

Danto, A (1964) "The Artworld," *Journal of Philosophy* 61.19:571–584.

Darwin, C. (1871) *The Descent of Man and Selection in Relation to Sex*. London, UK: John Murray.

Davies, D. (2013) "Dancing around the Issues: Prospects for an Empirically Grounded Philosophy of Dance," *Journal of Aesthetics and Art Criticism* 7.2:195–202.

Davies, D. (2014) "'This Is Your Brain on Art': What Can Philosophy of Art Learn from Neuroscience?" in *Aesthetics and the Sciences of the Mind*, pp. 57–74, ed. G. Currie, M. Kieran, A. Meskin and J. Robson. Oxford, UK: Oxford University Press.

Davies, S. (1991) *Definitions of Art*. Ithaca, NY: Cornell University Press.

Davies, S. (2002) "Definitions of Art," in *The Routledge Companion to Aesthetics*, pp. 169–179, ed. B. Gaut and D. M. Lopes. New York, NY: Routledge Press.

Davies, S. (2012) *The Artful Species: Aesthetics, Art and Evolution*. Oxford, UK: Oxford University Press.

Davies, S. (2015) "Defining Art and Artworlds," *Journal of Aesthetics and Art Criticism* 73.4:375–384.

Dickie, G. (1962) "Is Psychology Relevant to Aesthetics?" *Philosophical Review* 71.3:285–302.

Dickie, G. (1974) *Art and the Aesthetic: An Institutional Analysis*. Ithaca, NY: Cornell University Press.

Dickie, G. (1997) *Art Circle: A Theory of Art*. Chicago, IL: Spectrum Press.

Dissanayake, E. (1990) *What Is Art For?* Seattle, WA: University of Washington Press.

Dissanayake, E. (1995) *Homo Aestheticus*. Seattle, WA: University of Washington Press.

Dorsch, F. (2014) "The Limits of Aesthetic Empiricism," in *Aesthetics and the Sciences of the Mind*, pp. 75–100, ed. G. Currie, M. Kieran, A. Meskin and J. Robson. Oxford, UK: Oxford University Press.

Dutton, D. (2005) "Aesthetics and Evolutionary Psychology," in *The Oxford Handbook of Aesthetics*, pp. 693–721, ed. J. Levinson. Oxford, UK: Oxford University Press.

Dutton, D. (2010) *The Art Instinct: Beauty, Pleasure and Human Evolution*. New York, NY: Bloomsbury Press.

Geertz, C. (1976) "Art As a Cultural System," *Comparative Literature* 91.6:1473–1499.

Goldman, A. H. (2012) "Hume," in *Aesthetics: The Key Thinkers*, pp. 48–60, ed. A. Giovannelli. London, UK: Continuum.

Hickok, G. (2009) "Eight Problems for the Mirror Neuron Theory of Action Understanding in Monkeys and Humans," *Journal of Cognitive Neuroscience* 21.7:1229–1244.

Janaway, C. (2002) "Plato," in *The Routledge Companion to Aesthetics*, pp. 3–13, ed. B. Gaut and D. M. Lopes. New York, NY: Routledge.

Jordania, J. (2011) *Why Do People Sing? Music in Human Evolution*. Tbilisi, Georgia: LOGOS Publishing.

Kant, I. (1978) *Critique of Judgement*, ed. J. D. Meredith. Oxford, UK: Oxford-Clarendon Press.

Kemp, G. (2012) "Benedetto Croce and Robin G. Collingwood," in *Aesthetics: The Key Thinkers*, pp. 100–112, ed. A. Giovannelli. London, UK: Continuum.

Kieran, M. (2011) "The Fragility of Aesthetic Knowledge: Aesthetic Psychology and Appreciative Virtues," in *The Aesthetic Mind*, pp. 32–43, ed. E. Schellekens and P. Goldie. Oxford, UK: Oxford University Press.

Koelsch, S. (2010) "Towards a Neural Basis of Music-Evoked Emotions," *Trends in Cognitive Science* 14.3:131–137.

Kozbelt, A. and J. C. Kaufman (2014) "Aesthetic Assessment," in *The Cambridge Handbook of the Psychology of Aesthetics and the Arts*, pp. 86–112, ed. P. P. L. Tinio and J. K. Smith. Cambridge, UK: Cambridge University Press.

Kristeller, P. O. (1951) "The Modern System of the Arts: A Study in the History of Aesthetics Part I," *Journal of the History of Ideas* 12.4:496–527.

Levinson, J. (1979) "Defining Art Historically," *British Journal of Aesthetics* 19.3:232–250.

Levinson, J. (2002) "The Irreducible Historicality of the Concept of Art," *British Journal of Aesthetics* 42.4:367–369.

Livingstone, M. (2002) *Vision and Art: The Biology of Seeing*. New York, NY: Abrams.

Locher, P. J. (2014) "Empirical Investigation of the Elements of Composition in Paintings: A Painting As a Stimulus," in *The Cambridge Handbook of the Psychology of Aesthetics and the Arts*, pp. 221–242, ed. P. P. L. Tinio and J. K. Smith. Cambridge, UK: Cambridge University Press.

Lumsden, C. J. and E. O. Wilson (1981) Genes, *Mind and Culture: The Coevolutionary Process*. Cambridge, MA: Harvard University Press.

Lopes, D. M. (2014) "Feckless Reason," in *Aesthetics and the Sciences of the Mind*, pp. 21–36, ed. G. Currie, M. Kieran, A. Meskin and J. Robson. Oxford, UK: Oxford University Press.

Margolis, E. and S. Laurence (2014) "Concepts," *Stanford Encyclopedia of Philosophy* (Spring 2014 Edition), Edward N. Zalta (ed.), https://plato .stanford.edu/archives/spr2014/entries/concepts/.

Martindale, C. and K. Moore (1988) "Priming, Prototypicality and Preference," *Journal of Experimental Psychology, Human Perception and Performance* 14.4:661–670.

McFee, G. (2011) *The Philosophical Aesthetics of Dance: Identity, Performance and Understanding*. Hampshire, UK: Dance Books.

McFee, G. (2013) "Defending Dualism: John Martin on Dance Appreciation," *Journal of Aesthetics and Art Criticism* 71.2:187–194.

McIntosh, R. P. (1988) *The Background of Ecology: Concept and Theory*. Cambridge, UK: Cambridge University Press.

McNeill, W. H. (1995) *Keeping Together in Time: Dance and Drill in Human History*. Cambridge, MA: Harvard University Press.

Miller, G. (2001) *The Mating Mind: How Sexual Choice Shaped the Evolution of Human Behavior*. New York, NY: Anchor Books.

Mithen, S. (2006) *The Singing Neanderthals: The Origins of Music, Language, Mind and Body*. Cambridge, MA: Harvard University Press.

Montero, B. G. (2013) "The Artist As Critic: Dance Training, Neuroscience, and Aesthetic Evaluation," *Journal of Aesthetics and Art Criticism* 71.2:169–176.

Nadal, M. and G. Gómez-Puerto (2014) "Evolutionary Approaches to the Arts and Aesthetics," in *The Cambridge Handbook of the Psychology of Aesthetics and the Arts*, pp. 167–194, ed. P. P. L. Tinio and J. K. Smith. Cambridge, UK: Cambridge University Press.

Odling-Smee, F. J., K. N. Laland and M. W. Feldman (2003) *Niche Construction: The Neglected Process in Evolution*. Princeton, NJ: Princeton University Press.

Overy, K. and I. Molnar-Szakacs (2009) "Being Together in Time: Musical Experience and the Mirror Neuron System," *Music Perception: An Interdisciplinary Journal* 26.5:489–504.

Palmer, S. E., K. B. Schloss and J. Sammartino (2014) "Hidden Knowledge in Aesthetic Judgments: Preferences for Color and Spatial Composition," in *Aesthetic Science: Connecting Minds, Brains and Experience*, pp. 189–222, ed. A. P. Shimamura and S. E. Palmer. Oxford, UK: Oxford University Press.

Papineau, D. (2015) "Naturalism," *Stanford Encyclopedia of Philosophy* (Fall 2015 Edition), Edward N. Zalta (ed.), http://plato.stanford.edu/archives/fall2015/entries/naturalism/.

Pappas, N. (2002) "Plato," in *The Routledge Companion to Aesthetics*, pp. 15–26, ed. B. Gaut and D. M. Lopes. New York, NY: Routledge.

Pfaff, J. A., L. Zanette, S. A. MacDougall-Shackleton and E. A. MacDougall-Shackleton (2007) "Song Repertoire Size Varies with HVC Volume and Indicative of Male Quality in Song Sparrows," *Proceedings of Biological Sciences* 274.1621:2035–2040.

Pinker, S. (1997) *How the Mind Works*. New York, NY: Norton and Company.

Plato. (1997) *Complete Works*, ed. J. M. Cooper, Indianapolis, IN: Hackett.

Prum, R. (2013) "Coevolutionary Aesthetics in Human and Biotic Artworlds," *Biology and Philosophy* 28.5:811–832.

Puta, B. (2015) "Proto-Art and Art: Art before Art," in *The Genesis of Creativity and the Origin of the Human Mind*, pp. 45–56, ed. B. Puta and V. Soukup. Prague, Czech Republic: Karolinum Press, Charles University.

Ramachandren, V. and W. Hirstein (1999) "The Science of Art: A Neurological Theory of Aesthetic Experience," *Journal of Consciousness Studies* 6.6–7:5–51.

Reber, R. (2014) "Processing Fluency, Aesthetic Pleasure and Culturally Shared Taste," in *Aesthetic Science: Connecting Minds, Brains and Experience*, pp. 223–249, ed. A. P. Shimamura and S. E. Palmer. Oxford, UK: Oxford University Press.

Richards, R. A. (2004) "A Fitness Model of Evaluation," *Journal of Aesthetics and Art Criticism* 62.3:263–275.

Richards, R. A. (2005a) "Evolutionary Naturalism and the Logical Structure of Valuation: The Other Side of Error Theory," *Cosmos and History: The Journal of Natural and Social Philosophy* 1.2:270–294.

Richards, R. A. (2005b) "Reply to Dickie," *Journal of Aesthetics and Art Criticism* 63.3:283–287.

Richards, R. A. (2012) "Sexual Selection," in The Cambridge Encyclopedia of Darwin and Evolutionary Thought, pp. 103–108, ed. M. Ruse. Cambridge, UK: Cambridge University Press.

Richards, R. A. (2017a) "Engineered Niches and Naturalized Aesthetics," *Journal of Aesthetics and Art Criticism* 75.4:465–477.

Richards, R. A. (2017b) "Evolutionary Naturalism and Valuation," in *Cambridge Handbook for Evolutionary Ethics*, pp. 129–142, ed. R. Richards and M. Ruse. Cambridge, UK: Cambridge University Press.

Robinson, J. (2005) *Deeper than Reason: The Emotions and Their Role in Literature, Music and Art*. Oxford, UK: Oxford University Press.

Sacks, O. (2007) *Musicophilia: Tales of Music and the Brain*. New York, NY: Alfred A. Knopf.

Sapolsky, R. M. (2017) *Behave: The Biology of Humans at Our Best and Worst*. New York, NY: Penguin Press.

Scaglion, R. (1993) "Giant Yams and Cycles of Sex, Warfare and Ritual," in *Portraits of Culture: Ethnographic Originals*, pp. 3–24, ed. M. Ember and R. C. Ember. Englewood Cliffs, NJ: Prentice Hall.

Schellekens, E. (2012) "Immanuel Kant," in *Aesthetics: The Key Thinkers*, pp. 61–74, ed. A. Giovannelli. London: UK: Continuum.

Searle, J. R. (2010) *Making the Social World: The Structure of Human Civilization*. Oxford, UK: Oxford University Press.

Shelley, J. (2015) "The Concept of the Aesthetic," *Stanford Encyclopedia of Philosophy* (Winter 2015 Edition), Edward N. Zalta (ed.), http://plato.stanford.edu/archives/win2015/entries/aesthetic-concept/.

Shimamura, A. P. (2014) "Toward a Science of Aesthetics: Issues and Ideas," in *Aesthetic Science: Connecting Minds, Brains and Experience*, pp. 3–28, ed. A. P. Shimamura and S. E. Palmer. Oxford, UK: Oxford University Press.

Silvia, P. (2014) "Human Emotions and Aesthetic Experience: An Overview of Empirical Aesthetics," in *Aesthetic Science: Connecting Minds, Brains and Experience*, pp. 250–275, ed. A. P. Shimamura and S. E. Palmer. Oxford, UK: Oxford University Press.

Smith, M. (2012) "Triangulating Aesthetic Experience," in *Aesthetic Science: Connecting, Minds, Brains, and Experience*, pp. 80–106, ed. A. P. Shimamura and S. E. Palmer. Oxford, UK: Oxford University Press.

Sober, E. and D. S. Wilson (1998) *Unto Others: The Evolution and Psychology of Unselfish Behavior*. Cambridge, MA: Harvard University Press.

Stauffer, R. C. (1957) "Haeckel, Darwin, and Ecology," *Quarterly Review of Biology* 32.2: 138–144.

Stecker, R. (1996) "Alien Objections to Historical Definitions of Art," *British Journal of Aesthetics* 36.3:305–308.

Stecker, R. (1997) *Artworks: Definition, Meaning, Value*. University Park, PA: Pennsylvania State University Press.

Sterelney, K. (2003) *Thought in a Hostile World: The Evolution of Human Cognition*. Bodmin, UK: Blackwell Publishing.

Tolstoy, L. (1899) "What Is Art?" in *What Is Art and Essays on Art*, trans. A. Maude. Reprinted Indianapolis, IN: Hackett, 1996.

Trehub, S. E. (2003) "Musical Predispositions in Infancy: An Update," in *The Cognitive Neuroscience of Music*, pp. 3–20, ed. I. Peretz and R. Zatorre. Oxford, UK: Oxford University Press.

Vartanian, O. (2014) "Empirical Aesthetics: Hindsight and Foresight," in *The Cambridge Handbook of the Psychology of Aesthetics and the Arts*, pp. 6–34, ed. P. P. L. Tinio and J. K. Smith. Cambridge, UK: Cambridge University Press.

Walton, K. (1990) *Mimesis As Make-Believe: On the Foundation of the Representational Arts*. Cambridge, MA: Harvard University Press.

Werker, J. F. and R. C. Tees (1984) "Cross-Language Speech Perception: Evidence for Perceptual Reorganization during the First Year of Life," *Infant Behavioral Development* 1.7:49–63.

White, R. (2003) *Prehistoric Art: The Symbolic Journey of Humankind*. New York, NY: Harry N. Abrams, Inc.

Wilson, E. O. (1998) *Consilience: The Unity of Knowledge*. New York, NY: Alfred A. Knopf.

Wilson, G. M. (1986) *Narration in Light: Studies in Cinematic Point of View*. Baltimore, MD: Johns Hopkins University Press.

Winerman, L. (2005) "The Mind's Mirror," Monitor on Psychology 36.9:48, American Psychological Association, www.apa.org/monitor/oct05/mirror.aspx.

Zeki, S. (2000) *Inner Vision: An Exploration of Art and the Brain*. Oxford, UK: Oxford University Press.

Acknowledgments

In memory of Willam F. Christensen, mentor and friend.

I have benefited from my associations and discussions with many people. My ballet mentor, Willam F. Christensen, may have had little formal education, but he was one of the most philosophically sophisticated people I have known. My partner in life and dance, Rita Snyder, a former professional symphony musician and professional dancer, has taught me much. I have also benefited greatly from discussions with my colleagues, Max Hocutt and Norvin Richards, as well as with my students, especially Brett Smith and Maria Gerasikova. Ben Kozuch and Nathan Ahlgrim provided feedback on Section 4, which was very helpful and for which I thank them. The basic idea for this Element was developed from my "Engineered Niches and Naturalized Aesthetics," *Journal of Aesthetics and Art Criticism* 75.4, 2017. That article benefited from the suggestions of the editors of that journal. Nonetheless, all errors are my own. I am forever grateful for the philosophical mentoring, friendship and support of Peter Achinstein, and I cannot thank Michael Ruse enough for his generous support and guidance. It has been a pleasure to work with Hilary Gaskin of Cambridge University Press.

Cambridge Elements $\equiv$

Philosophy of Biology

Grant Ramsey

KU Leuven

Grant Ramsey is a BOFZAP Research Professor at the Institute of Philosophy, KU Leuven, Belgium. His work centers on philosophical problems at the foundation of evolutionary biology. He has been awarded the Popper Prize twice for this work in this area. He also publishes in the philosophy of animal behavior, human nature, and the moral emotions. He runs the Ramsey Lab (theramseylab.org), a highly collaborative research group focused on issues in the philosophy of the life sciences.

Michael Ruse

Florida State University

Michael Ruse is the Lucyle T. Werkmeister Professor of Philosophy and the Director of the Program in the History and Philosophy of Science at Florida State University. He is Professor Emeritus at the University of Guelph, in Ontario, Canada. He is a former Guggenheim fellow and Gifford lecturer. He is the author or editor of over sixty books, most recently *Darwinism as Religion: What Literature Tells Us about Evolution; On Purpose; The Problem of War: Darwinism, Christianity, and their Battle to Understand Human Conflict*; and *A Meaning to Life*.

About the Series

This Cambridge Elements series provides concise and structured introductions to all of the central topics in the philosophy of biology. Contributors to the series are cutting-edge researchers who offer balanced, comprehensive coverage of multiple perspectives, while also developing new ideas and arguments from a unique viewpoint.

Cambridge Elements ≡

Philosophy of Biology

Elements in the Series

Printed in the United States
By Bookmasters